T0223771

Automatic
Flight Control Systems

Synthesis Lectures on Mechanical Engineering

Synthesis Lectures on Mechanical Engineering series publishes 60–150 page publications pertaining to this diverse discipline of mechanical engineering. The series presents Lectures written for an audience of researchers, industry engineers, undergraduate and graduate students.

Additional Synthesis series will be developed covering key areas within mechanical engineering.

Automatic Flight Control Systems
Mohammad Sadraey
2019

Bifurcation Dynamics of a Damped Parametric Pendulum
Yu Guo and Albert C.J. Luo
2019

Reliability-Based Mechanical Design, Volume 1: Component under Cyclic Load and Dimension Design with Required Reliability
Xiaobin Le
2019

Reliability-Based Mechanical Design, Volume 1: Component under Static Load
Xiaobin Le
2019

Solving Practical Engineering Mechanics Problems: Advanced Kinetics
Sayavur I. Bakhtiyarov
2019

Natural Corrosion Inhibitors
Shima Ghanavati Nasab, Mehdi Javaheran Yazd, Abolfazl Semnani, Homa Kahkesh, Navid Rabiee, Mohammad Rabiee, Mojtaba Bagherzadeh
2019

Unmanned Aircraft Design: A Review of Fundamentals
Mohammad Sadraey
2017

Introduction to Refrigeration and Air Conditioning Systems: Theory and Applications
Allan Kirkpatrick
2017

Resistance Spot Welding: Fundamentals and Applications for the Automotive Industry
Menachem Kimchi and David H. Phillips
2017

MEMS Barometers Toward Vertical Position Detecton: Background Theory, System
Prototyping, and Measurement Analysis
Dimosthenis E. Bolanakis
2017

Engineering Finite Element Analysis
Ramana M. Pidaparti
2017

Automatic Flight Control Systems
Mohammad Sadraey

ISBN: 978-3-031-79648-7   paperback
ISBN: 978-3-031-79649-4   ebook
ISBN: 978-3-031-79651-7   hardcover

DOI 10.1007/978-3-031-79649-4

A Publication in the Springer series
SYNTHESIS LECTURES ON MECHANICAL ENGINEERING

Lecture #23
Series ISSN
Print 2573-3168   Electronic 2573-3176

Automatic Flight Control Systems

Mohammad Sadraey
Southern New Hampshire University

SYNTHESIS LECTURES ON MECHANICAL ENGINEERING #23

ABSTRACT

This book provides readers with a design approach to the automatic flight control systems (AFCS). The AFCS is the primary on-board tool for long flight operations, and is the foundation for the airspace modernization initiatives. In this text, AFCS and autopilot are employed interchangeably. It presents fundamentals of AFCS/autopilot, including primary subsystems, dynamic modeling, AFCS categories/functions/modes, servos/actuators, measurement devices, requirements, functional block diagrams, design techniques, and control laws.

The book consists of six chapters. The first two chapters cover the fundamentals of AFCS and closed-loop control systems in manned and unmanned aircraft. The last four chapters present features of Attitude control systems (Hold functions), Flight path control systems (Navigation functions), Stability augmentation systems, and Command augmentation systems, respectively.

KEYWORDS

automatic flight control systems, autopilot, attitude control systems, flight path control systems, stability augmentation systems, command augmentation systems

Contents

Preface

This books deals with the fundamentals of automatic flight control systems (AFCS). The AFCS is the primary on-board tool for long flight operations, and is the foundation for the airspace modernization initiatives. In this text, AFCS and autopilot are employed interchangeably. Nowadays, technological advances in wireless communication and micro-electro-mechanical systems make it possible to use inexpensive small autopilots in unmanned aircraft. This book presents fundamentals of AFCS/autopilot, including primary subsystems, dynamic modeling, and AFCS categories and modes, measurement devices, servos/actuators, requirements, functional block diagrams, and control laws.

In general, a closed-loop control system tends to provide four functions: (1) Regulating; (2) Tracking; (3) Stabilizing; (4) Improve the plant response. In flight control, the regulating function is referred to as the hold function, and such a system is referred to as the attitude control system. This topic with six applications is presented in Chapter 3. However, the tracking function is referred to as the navigation function, and such a system is referred to as the flight path control systems. Characteristics of eleven navigation functions of an AFCS are discussed in Chapter 4.

Negative feedback(s) can also stabilize an unstable vehicle. This function is provided by a system referred to as the stability augmentation system. Features of six stability augmentation systems are provided in Chapter 5. Moreover, when negative feedback(s) is employed to improve an unsatisfactory system-response, the function is called the command augmentation function. The details of four command augmentation systems are discussed in Chapter 6.

A significant element in any unmanned aircraft, fighters, and large manned aircraft is the AFCS. The AFCS is a rigorous topic; the reader must be aware of the following math and engineering fundamentals: differential equations, linear algebra, calculus, flight dynamics, aerodynamics, and control systems.

The AFCSs are under constant progress and growth; every year new AFCS modes are being designed and introduced. To demonstrate the modern applications, the features of AFCS of a number of general aviation, transport and fighter aircraft, and unmanned aerial vehicles have been briefly introduced.

This book is intended to be used in courses in flight dynamics, flight control systems, and unmanned aerial vehicles (UAV). As a main text. This author hopes that such intention is indeed achieved in the present book. Due to limited number of pages, a number of prerequisite topics such as "fundamentals of control systems," "flying qualities," and "dynamic modeling" are briefly covered. Moreover, no in-chapter solved examples and end-of-chapter problems are developed. I hope the reader enjoys reading this book, and learns the applications of closed-loop control in aerial vehicles flight. I wish to thank Morgan & Claypool Publishers and particularly Paul

Petralia, Executive Editor in engineering, for his support during this period, and Sara Kreisman for copy-editing the manuscript.

Mohammad Sadraey
December 2019

C H A P T E R 1

Fundamentals of Automatic Flight Control Systems

1.1 INTRODUCTION

The evolution of modern aircraft in 1950s created a need for automatic-pilot control systems. The original intent of the Automatic Flight Control Systems (AFCS) was to offload routine tasks from the pilot and enhance fuel efficiencies for enroute flight operations. The AFCS is the primary on-board tool for long flight operations, and is the foundation for the airspace modernization initiatives. To reduce pilot workload—particularly on long flights—modern transport aircraft are all equipped with AFCS.

In the 1970s, advances in AFCS allowed the fighter aircraft General Dynamics F-16 Fighting Falcon to be designed for fly-by-wire control and "relaxed static stability." Fighter aircraft AFCSs are frequently designed to provide the pilot control over pitch rate at low speed, and normal acceleration (n_z) at high speed. The AFCS optimizes the airplane's performance capability and accuracy along the planned route. Flight automation is the future, with of course, manual flying interference in extreme situations. The task of designing AFCS is considerable, and involves much analysis.

In this text, AFCS and autopilot are employed interchangeably. Autopilot is the integrated software and hardware that serve three functions: (1) Control, (2) Navigation, and (3) Guidance. In a typical autopilot, three laws are governing simultaneously in three subsystems.

1. Control system → Control law

2. Guidance system → Guidance law

3. Navigation system → Navigation law

In the design of an autopilot, all three laws need to be selected/designed. The autopilot has the responsibility to: (1) stabilize under-damped or unstable modes and (2) accurately track commands generated by the guidance system.

An autopilot is capable of implementing many very time intensive tasks, which helps the pilot focus on the overall status of the flight. Tasks include maintaining an assigned altitude/airspeed/heading, climbing or descending to an assigned altitude, turning, intercepting a course, guiding the aircraft between waypoints that make up a programmed route, and flying a precision or non-precision approach. Moreover, when the autopilot keeps the aircraft on the programmed heading/course/altitude/airspeed, the pilot have free time to make the necessary changes to the flight plan.

The most important subsystem in an unmanned aerial vehicle (UAV), compared with a manned aircraft is the autopilot, since there is no human pilot in a UAV. The autopilot is a vital and required subsystem within unmanned aerial vehicles. An autopilot is an electro-mechanical device that must be capable of accomplishing all types of controlling functions including automatic take-off, flying toward the target destination, perform mission operations (e.g., surveillance), and automatic landing. The autopilot has the responsibility (i.e., functions) to: (1) stabilize the UAV; (2) track commands; (3) guide the UAV; and (4) navigate.

In general, a closed-loop (i.e., negative feedback) control system tends to provide four functions: (1) Regulating; (2) Tracking; (3) Stabilizing; and (4) Improve the plant response. The regulating function here is referred to as the hold function (e.g., the Learjet 75 aircraft), and such system is referred to as the attitude control system (see Chapter 3). However, the tracking function is referred to as the navigation function (e.g., the Boeing 767 aircraft), and such system is referred to as the flight path control systems (see Chapter 4). Moreover, a negative feedback can also stabilize an unstable plant (e.g., the B-2 aircraft). This function is referred to as the stability augmentation function (see Chapter 5). When a negative feedback is employed to improve an unsatisfactory system-response (e.g., the F-14 aircraft), the function is called the control augmentation function (see Chapter 6).

Two primary goals of an AFCS are: (1) closed-loop stability and (2) acceptable performance (time responses). Once the design objectives have been selected, the controller is explicitly finalized such that (1) closed-loop stability is guaranteed and (2) response to a step-input is satisfactory.

Nowadays, technological advances in wireless communication and micro electromechanical systems, make it possible to use inexpensive small autopilots. One of the most valuable benefits of using the autopilot is delegating the constant task of manipulating the control surfaces and engine throttle. This benefit allows the pilot more time to manage and observe the entire flight situation. This chapter presents fundamentals of AFCS/autopilot, including primary subsystems, dynamic modeling, and AFCS categories and modes.

1.2   ELEMENTS OF AN AFCS

1.2.1   RELATIONS BETWEEN AFCS AND HUMAN PILOT

The general relations between AFCS, human pilot, and the flight parameters is shown in Figure 1.1. There is one flight mission, one aircraft dynamics, but two sources of implementation. Both human pilot and AFCS may apply input the aircraft; simultaneously or individually. In most flight cases, only one of them should be controlling the aircraft. However, there are AFCS modes (e.g., yaw damper) that can be engaged while the human pilot is controlling the aircraft. The pilot has the authority to engage or disengage (on/off switch) any AFCS mode that he/she desires.

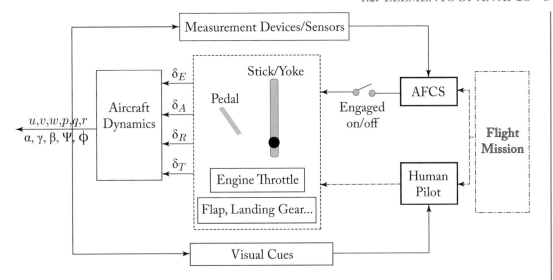

Figure 1.1: General relations between AFCS, human pilot, and the flight parameters.

The human pilot and AFCS each has a unique set of measurement devices, plus many common sensors. For both pilot and the AFCS, there are sensors such as three attitude gyros, three rate gyros, three magnetometers, IMU, airspeed meter, GPS, radar, transmitter, receiver, and altimeter. For the case of AFCS, the sensor signals are transmitted to computer to calculate the command signal. However, for the case of the pilot, the measures values are illustrated through gages and displays (i.e., visual cues). A pilot utilize this information to determine their position and navigate the aircraft to their destination. Autopilot modes may be disconnected by the mode buttons on the autopilot control panel.

The aircraft control may be implemented via mechanical, hydraulic, and electrical components and control wires. In large transport and fighter aircraft, the control surfaces are fully power driven, there is no force or motion feedback to the pilot's control stick. This is refered to as an *irreversible* control system, and bob weights and springs should be added to the control stick/yoke to provide some "feel" to the pilot. The control surfaces are driven by a hydraulic servomechanism and electrohydraulic valves. The stick/yoke and rudder pedals are linked to the servos/actuators by hydro-mechanical links.

1.2.2   PRIMARY SUBSYSTEMS OF AN AFCS

The primary functions of an autopilot are to: (1) accurately track commands generated by the command system in parallel with the guidance system; (2) guide the UAV to follow the trajectory; (3) stabilize under-damped or unstable modes; and (4) determine UAV coordinates (i.e., navigation). Therefore, an autopilot consists of: (1) command subsystem; (2) control subsystem; (3) guidance subsystem; and (4) navigation subsystem. These four subsystems must be

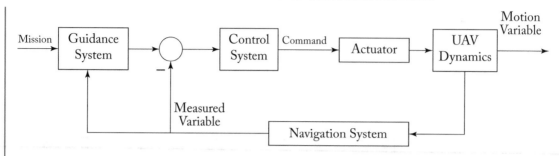

Figure 1.2: Control, guidance, and navigation systems in an autopilot.

designed simultaneously to satisfy the UAV design requirements. These subsystems include particular hardware (i.e., physical equipment), as well as the separate software (code) section. All required equipment must be designed and built, while all codes need to be written, compiled, and uploaded into the microcontroller.

In a conventional autopilot, three laws are simultaneously governing the three subsystems: (1) control system through a control law; (2) guidance system via a guidance law; and (3) navigation system through a navigation law. In the design of an autopilot, all three laws need to be developed/designed. The design of the control law is at the heart of autopilot design process. The relation between the control system, the guidance system, and the navigation system is shown in Figure 1.2. This figure is not illustrating the command system, it will be covered later.

In analyzing any AFCS mode, two basic information are needed: (1) type of controller which should be employed to control the selected flight variable; and (2) type and number of feedbacks and respective sensors.

1.2.3   AUTONOMY CLASSIFICATION

A typical full flight consists of the following phases: (1) Ground taxi, including ground collision avoidance; (2) Take-off; (3) Climb; (4) En-route cruise; (5) Turn and maneuver; (6) Descent; (7) Landing; (8) Ground operation at the destination; and (9) Handling of emergencies in any of these sectors. Autonomy is defined with respect to every flight phase, and with various levels. Moreover, some high-level autonomy such as detect-and-avoid, and fault monitoring will be addressed in this text.

In general, there are four ways for piloting a UAV: (1) remote control; (2) autopilot-assisted control (i.e., automated); (3) semi-full autonomy; and (4) full autonomy. Under full autonomous control, the reality is that the on-board computer is in control not a human operator. A minimal autopilot system includes attitude sensors and onboard processor. Reference [41] divides autonomy into 11 (from 0–10) levels as a gradual increase of Guidance-Navigation-Control (GNC) capabilities: (0) Remote Control; (1) Automatic Flight Control; (2) External System Independence Navigation; (3) Fault/Event Adaptive; (4) Real Time Obstacle/Event Detection and Path Planning; (5) Real Time Cooperative Navigation and Path Planning; (6) Dynamic Mission Planning; (7) Real time Collaborative Mission Planning; (8) Situational Awareness and Cognizance; (9) Swarm Cognizance and Group Decision Making; and (10) Fully Autonomous.

In the automated or automatic system, in response to feedbacks from one or more sensors, the aircraft is programmed to logically follow a pre-defined set of rules in order to provide an output. Knowing the set of rules under which it is operating means that all possible output are predictable. A modern autopilot allows the vehicle to fly on a programmed flight paths without human interference for almost all the mission, without an operator doing anything other than monitoring its operation.

Currently, there is no consensus for the definition of autonomy in UAV community. In this book, "autonomy" is the ability of an agent to carry out a mission in an independent fashion without requiring human intervention. Autonomy should include kind of artificial intelligence, since the decision-making is performed by the autopilot. An autonomous vehicle is capable of understanding higher level intents and directions. Such a vehicle is able to take appropriate action to bring about a desired state/trajectory, based on this understanding and its perception of the environment, it is capable of deciding a course of action, from a number of alternatives, without depending on human oversight and control, although these may still be present (for monitoring). Although the overall activity of an autonomous UAV is predictable, individual actions may not be. An autonomous UAV is able to monitor and assess its status (e.g., altitude, airspeed) and configuration (e.g., flap deflection), and command and control assets onboard the vehicle.

The core components of autonomy are command, control, navigation and guidance. Higher levels of autonomy, which reduce operator workload, include (in increasing order) sense-and-avoid, fault monitoring, intelligent flight planning, and reconfiguration. An autonomous behavior includes observe, orient, decide, and act. The aviation industry objective is that eventually autonomous UAV will be able to operate without human intervention across all flight sectors. Such objective requires advances in various technologies including guidance system, navigation system, control system, sensors, avionics, communication systems, infrastructures, software, and microprocessors.

1.3   FLIGHT DYNAMICS

1.3.1   DYNAMICS MODELING

The design of AFCS requires a variety of mathematical equations and technical information including the UAV dynamics. The first step to analysis and design an autopilot is to describe the dynamic behavior of the vehicle in a mathematical language. The quantitative mathematical description of a physical system is known as dynamic modeling. There are several ways for mathematical descriptions. The most widely used method is the differential equation.

Description of the behavior and components of a dynamic system with a mathematical language is referred to as the dynamic modeling. The dynamic behavior (i.e., UAV dynamics) of a flight vehicle is based on Newton's second law: when a force is applied on an object, the acceleration of the object is directly proportional to the magnitude of the force, in the same direction as the force, and inversely proportional to the mass of the object.

Once a physical system (e.g., aircraft) has been described by a set of mathematical equations, they are manipulated to achieve an appropriate mathematical format. There are mainly two techniques to model a dynamic system: (1) transfer function and (2) state space representation. The first one is described in s-domain (frequency); the second one is presented in time domain. A dynamic system may also by represented pictorially; this method is using a block diagram. An UAV is a dynamic system and its dynamic behavior will be modeled by either way. Using an aircraft dynamic model, one is able to design AFCS to satisfy the design requirements.

One form of the mathematical model of a SISO plant is the transfer function. The plants are dynamic in nature, and the mathematical models are usually differential equations. If these equations are linearized, then the Laplace transform can be utilized to develop transfer function. In general, the format of a transfer function is the ratio of two polynomials in s:

$$F(s) = \frac{b_1 s^m + b_2 s^{m-1} + \ldots + b_{m+1}}{s^n + a_1 s^{n-1} + \ldots + a_n}. \tag{1.1}$$

A proper second-order system is modeled as

$$F(s) = \frac{N(s)}{as^2 + bs + c}, \tag{1.2}$$

where $N(s)$ represents the numerator; which in this case, is a first-order polynomial. The standard form for a pure second-order dynamic system is a function of two parameters of damping ratio (ξ) and natural frequency (ω_n) as:

$$F(s) = \frac{k}{s^2 + 2\xi\omega_n s + \omega_n^2}. \tag{1.3}$$

The desired damping ratio and natural frequency are employed in the design process to satisfy the control system design requirements. The denominator of a transfer function is referred to as the characteristic polynomial/equation, since it reveals the stability characteristics of the dynamic system. In a first-order system (purely exponential), the main performance parameter is the time constant.

The transfer function belongs to era of classical control and only is able to cover a single-input-single-output (SISO) system. Using classical control theory, the designer is forced to take a one-loop-at-a-time approach. However, for a Multi-Input-Multi-Output (MIMO) system, a matrix format referred to as the state space representation is employed. This mathematical model consists of series of first-order linear differential equations and a series of linear algebraic equations:

$$\dot{x} = Ax + Bu$$
$$y = Cx + Du,$$
(1.4)

where the A, B, C, and D are matrices, x is state variable, u is the control (input) variable, and y is the output variable.

The control system design process requires the aircraft mathematical model as the basis for the design. The aircraft modeling process basically consists of: (1) dynamics modeling; (2) aerodynamics modeling; (3) engine modeling; and (4) structural modeling. An aircraft is basically a nonlinear system. Furthermore, its dynamics and equations of motion are also nonlinear. The aircraft dynamic model is could be linearized or decoupled. The nonlinear coupled equations of motion are the most complete dynamic model. The flat-Earth vector equations of motion are often used, and when they are expanded the standard 6-DOF equations used for aircraft control design and flight simulation are obtained. Other aircraft dynamic models are: (1) linear decoupled equations of motion; (2) hybrid coordinates; (3) linear coupled equations of motion; and (4) nonlinear decoupled equations of motion.

The aerodynamic forces and moments (aerodynamics model) of the complete aircraft are defined in terms of dynamic pressure, aircraft geometry, and dimensionless aerodynamic coefficients. The aerodynamic coefficients are assumed to be the linear functions of state variables and control inputs. These forces and moments are used in equations of motion (dynamic model) as part of control system design, as well as the flight simulation.

The propulsion model is based on the propulsion system powering the aircraft (e.g., prop-driven or jet engine). In case of a jet engine, the engine thrust (T) is modeled in terms of throttle setting (δ_T). In case of a prop-driven engine, the engine power is a function of required engine thrust, aircraft speed, and propeller efficiency. Several control system design techniques may be found in the literature. The choice of the type of control system depends on a variety of factors including the system's characteristics and the design requirements. The heart of the control system is the controller. There are several textbooks and papers regarding controller design in the literature; references [38] and [39] present several controller design techniques.

The real aircraft behavior is nonlinear and has uncertainties. It is also important to note that all measurement devices (including gyros and accelerometers) have some kind of noise that must be filtered. It is well known that the atmosphere is a dynamic system that produces lots of disturbances throughout the aircraft's flight. Finally, since fuel is expensive and limited, and actuators have dynamic limitations, optimization is necessary in control system design. Therefore, it turns out that only a few design techniques, such as robust nonlinear control, are able to satisfy all safety, cost, and performance requirements. In order to select the best controller technique, one must utilize a trade-off study and compare the advantages and disadvantage of all candidate controllers.

Virtually all dynamic systems are nonlinear; yet an overwhelming majority of operational control laws have been designed as if their dynamic systems were linear and time-invariant. As long as the quantitative differences in response are minimal (or at least acceptable in some practical sense), the linear time invariant model facilitates the control system design process. This is because of the direct manner in which response attributes can be associated with model parameters.

A small error in modeling, a small error in the control system design, or a small error in the simulation may each result in problems in flight, in the worst case might even result in the loss of an unmanned aerial vehicle. As long as the dynamic effects of parameter variations are slow in comparison to state variation, control design can be based on an ensemble of time-invariant dynamic models. Fast parameters may be indistinguishable from state components, in which case the parameters should be included in an augmented state vector for estimation. The aircraft field is one where extensive use is made of modeling and simulation technologies.

The complete aircraft systems and dynamics model incorporates different subsystem models (e.g., aerodynamics, structures, propulsion, and control subsystems) that have interdependent responses to any input. These subsystems also interact with the other subsystems. The dynamic modeling of an aircraft is at the heart of its simulation. The response of an aerial vehicle system to any input, including commands or disturbances (e.g., wind gusts), can be modeled by a system of ordinary differential equations (i.e., the equations of motion). Dealing with the nonlinear, fully coupled differential equations of motion is not an easy task.

The dynamics of an aircraft can be modeled in different ways. The equations of motion take several forms including: (1) nonlinear fully coupled; (2) nonlinear semi-coupled; (3) nonlinear decoupled; (4) nonlinear reformulated; (5) linear coupled; (6) linear decoupled; and (7) linear time-variant.

To develop a computer simulation to evaluate the performance of an aerial vehicle (manned or unmanned) including its control system, we have to invariably use a nonlinear fully coupled model. In order to design a control system, one of the above models are utilized. Each of these models has advantages and disadvantages. These include precision, accuracy, complexity, and credibility. The use of flight simulation tools to reduce risk and flight testing for an aerial vehicle system reduces the overall program schedule.

1.3.2   FLIGHT FUNDAMENTAL GOVERNING EQUATIONS

Newton's Second Law states that the time derivatives of linear momentum ($m \cdot V$) and angular momentum ($I \cdot \omega$) are equal to the externally applied forces (F) and moments (M), respectively:

$$\sum F = \frac{d}{dt}(m \cdot V_T) \tag{1.5}$$

$$\sum M = \frac{d}{dt}(I \cdot \omega), \tag{1.6}$$

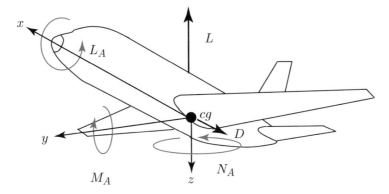

Figure 1.3: Coordinate axes, aerodynamic forces, and moments.

where m, V_T, I, and ω are aircraft mass, total airspeed, aircraft mass moment of inertia, and angular speed, respectively. As Figure 1.3 illustrates, an aircraft has three axes (x, y, and z), hence the vehicle has six degree-of-freedom (DOF), three linear displacement along three axes, and three angular displacement about the three axes. In contrast, there are three linear velocities (U, V, and W) and three angular velocities (P, Q, and R). Three angular velocities represent roll rate about x-axis (P), pitch rate about y-axis (Q), and yaw rate about z-axis (R).

Therefore, there are three groups of forces, including aerodynamic forces (F_A) and thrust forces (F_T), and three aerodynamic moments. The aerodynamic forces D, Y, and L, are the drag, side-force, and lift, and L_A, M_A, N_A, are the aerodynamic moments. The aerodynamic forces and moments are reviewed in the next section. Applying forces, moments, and velocities into Equations (1.5) and (1.6), the following standard body-axis six first-order nonlinear differential equations are obtained [2].

Force equations:

$$m\left(\dot{U} - VR + WQ\right) = mg_x + F_{A_x} + F_{T_x} \tag{1.7}$$

$$m\left(\dot{V} + UR - WP\right) = mg_y + F_{A_y} + F_{T_y} \tag{1.8}$$

$$m\left(\dot{W} - UQ + VP\right) = mg_z + F_{A_z} + F_{T_z}. \tag{1.9}$$

Moment equations:

$$\dot{P}I_{xx} - \dot{R}I_{xz} - PQI_{xz} + \left(I_{zz} - I_{yy}\right)RQ = L_A + L_T \tag{1.10}$$

$$\dot{Q}I_{yy} + \left(I_{xx} - I_{zz}\right)PR + I_{xz}\left(P^2 - R^2\right) = M_A + M_T \tag{1.11}$$

$$\dot{R}I_{zz} - \dot{P}I_{xz} + \left(I_{yy} - I_{xx}\right)PQ + I_{xz}QR = N_A + N_T. \tag{1.12}$$

Equations (1.7)–(1.9) govern linear motion along x, y, and z axes, and constitutes force equations of the equations of motion. In contrast, Equations (1.10)–(1.12) govern angular motion about x, y, and z axes, and constitutes the moment equations of the equations of motion.

Many aircraft are symmetric about x-z plane, so the cross-product of inertia (I_{xz}) may be assumed zero. Furthermore, the thrust moments (L_T, M_T, N_T) are also negligible. Applying these two assumptions provides a simpler versions of the moment equations as:

$$\dot{P} = \frac{I_{yy} - I_{zz}}{I_{xx}} QR + \frac{L_A}{I_{xx}} \tag{1.13}$$

$$\dot{Q} = \frac{I_{zz} - I_{xx}}{I_{yy}} RP + \frac{M_A}{I_{yy}} \tag{1.14}$$

$$\dot{R} = \frac{I_{xx} - I_{yy}}{I_{zz}} PQ + \frac{N_A}{I_{zz}}. \tag{1.15}$$

These moment equations are sometimes referred to as the Euler's equations. The attitude angles are computed by integrating the angular rates, not from trigonometric functions. In order to develop control laws, Equations (1.7)–(1.12) are often converted to other forms such as: (1) transfer functions and (2) state space representation. When more than one motion variable must be fed back to a number of controllers (i.e., control surfaces); mathematical complication arises in the derivation of transfer functions.

1.3.3   NONLINEAR FULLY COUPLED EQUATIONS OF MOTION

The standard body-axis nonlinear fully coupled equations of motion include three force—and three moment—first-order differential equations (in state-space model) as follows (Ref. [1]):

$$\dot{U} = RV - WQ - g \sin \theta + \frac{1}{m} [-D + T \cos \alpha] \tag{1.16}$$

$$\dot{V} = -UR + WP + g \sin \phi \cos \theta + \frac{1}{m} [Y + T \cos \alpha \sin \beta] \tag{1.17}$$

$$\dot{W} = UQ - VP + g \cos \phi \cos \theta + \frac{1}{m} [-L - T \sin \alpha] \tag{1.18}$$

$$\dot{P} = (c_1 R + c_2 P) Q + c_3 (L_A + L_T) + c_4 (N_A + N_T) \tag{1.19}$$

$$\dot{Q} = c_5 PR + c_6 (P^2 - R^2) + c_7 M \tag{1.20}$$

$$\dot{R} = (c_8 P - c_2 R) Q + c_4 (L_A + L_T) + c_9 (N_A + N_T). \tag{1.21}$$

In the above equations, the parameters c_i are functions of the aircraft moments of inertia, and can be calculated from the formula introduced in Ref. [1]. The parameters U, V, W are the linear velocity components, and P, Q, R are the corresponding angular rates. The aerodynamic forces D, Y, and L, are the drag, side-force, and lift, and L_A, M_A, N_A, are the aerodynamic moments. The variables α, β, ϕ, and θ are angle of attack, sideslip angle, bank angle, and pitch angle, respectively. The motions of a UAV are: longitudinal, lateral, and directional.

1.3.4 LINEAR DECOUPLED EQUATIONS OF MOTION

Decoupling is based on the assumption that the longitudinal motion is independent of the lateral-directional motion. When decoupling (no coupling effects) and linearization techniques are applied simultaneously to Equations (1.3)–(1.8), the state-space equations [5] are split into two groups, each having four states, two inputs, and four outputs.

A fundamental point when applying this form is that they are reliable only in the vicinity of the initial *trim point*. The validity of these equations is conversely related to the distance from the trim point. As the flight condition get further from trim point, the validity of the result is reduced. The linear decoupled equations of motion are:

$$\dot{x} = Ax + Bu$$
$$y = Cx + Du,$$
(1.22)

where the A, B, C, and D matrices are presented below.

a. Longitudinal state-space model

$$
\begin{bmatrix} \dot{u} \\ \dot{w} \\ \dot{q} \\ \dot{\theta} \end{bmatrix}
=
\begin{bmatrix}
X_u & X_w & 0 & -g \\
Z_u & Z_w & u_o & 0 \\
M_u & M_w & M_q & 0 \\
0 & 0 & 1 & 0
\end{bmatrix}
\begin{bmatrix} u \\ w \\ q \\ \theta \end{bmatrix}
+
\begin{bmatrix}
X_{\delta_E} & X_{\delta_T} \\
Z_{\delta_E} & Z_{\delta_T} \\
M_{\delta_E} & M_{\delta_T} \\
0 & 0
\end{bmatrix}
\begin{bmatrix} \delta_E \\ \delta_T \end{bmatrix}.
$$
(1.23)

b. Lateral-directional state-space model

$$
\begin{bmatrix} \dot{\beta} \\ \dot{p} \\ \dot{r} \\ \dot{\phi} \end{bmatrix}
=
\begin{bmatrix}
\dfrac{Y_\beta}{u_o} & \dfrac{Y_p}{u_o} & -1 + \dfrac{Y_r}{u_o} & \dfrac{g \cos \theta}{u_o} \\
L_\beta & L_p & L_r & 0 \\
N_\beta & N_p & N_r & 0 \\
0 & 1 & 0 & 0
\end{bmatrix}
\begin{bmatrix} \beta \\ p \\ r \\ \phi \end{bmatrix}
+
\begin{bmatrix}
0 & \dfrac{Y_{\delta_A}}{u_o} \\
L_{\delta_A} & L_{\delta_R} \\
N_{\delta_A} & N_{\delta_R} \\
0 & 0
\end{bmatrix}
\begin{bmatrix} \delta_A \\ \delta_R \end{bmatrix}.
$$
(1.24)

In other words, the longitudinal motion is independent of the lateral-directional motion. In both longitudinal (Equation (1.23)) and lateral-directional (Equation (1.24)) models, the C is a 4×4 identity matrix, and D is a 4×2 matrix where all elements are zero.

1.4 AFCS CATEGORIES AND MODES

The fundamental function of an automatic flight control system is to control the flight variables to ensure the aircraft is flying in the predetermined trajectory. In a UAV, since there is no human pilot involved, an autopilot is a device that must be capable of accomplishing all types of controlling functions including automatic take-off, flying toward the target destination, and automatic landing.

The simplest autopilot for small single engine prop-driven aircraft is called "single axis" which controls the ailerons only. It can hold the wing level, and can stabilize the airplane in terms of rolling motion. This autopilot can also maintain a heading or track a set of GPS coordinate waypoints. The next level of autopilot can control both roll and pitch via aileron and elevator. This allows the autopilot to maintain altitude or climb rate.

Small GA aircraft like Cessna Citation generally have less advanced AFCSs than large transport aircraft made by Airbus or Boeing. The overall architecture and redundancy management concepts of autopilot, flight director and AFCS of the transport aircraft MD-11 is introduced in [26]. Important modes include: automatic landing, wind shear function, longitudinal stability augmentation, roll control wheel steering, speed envelope protection via auto throttle, yawdamper, turn coordination, elevator load feel and flap limiting control, altitude alert, stall warning with stick shaker, automatic slat extension; automatic ground spoilers, wheel spin-up and horizontal stabilizer in motion detection, automatic throttle and engine trim, take-off, cruise, and CAT II approach.

In general, there are five categories/functions for AFCSs: (1) stability augmentation; (2) hold functions; (3) navigation functions; (4) command augmentation; and (5) combined category. These features are often referred to as autopilot modes.

The features of first four categories are briefly explained in the following sections and described in details in Chapters 3–6. In the fifth category (i.e., combined), various functions are combined for a complex mission. For instance, the autopilot of General Atomics Reaper MQ-9B (a Certifiable version of its Predator B) had been configured in an Intelligence, Surveillance, and Reconnaissance (ISR) for the entire mission. This UAV, on May 25, 2017, had an endurance of 48.2 hours.

1.4.1   STABILITY AUGMENTATION SYSTEMS

For a lightly stable aircraft, the autopilot must provide the augmented stability. As the name implies, a stability augmentation system (SAS) is to augment the stability of an open-loop plant. It improves the aircraft stability, and even stabilizes an unstable aircraft. The SAS can also be simultaneously employed with manual control (Ground-system/operator command). Due to three aircraft axes, there are three stability augmentation systems, namely: (1) roll damper, (2) yaw damper, and (3) pitch damper.

The widening flight envelope (e.g., Figure 2.5) creates a need to augment the stability of the aircraft dynamics over some parts of the flight envelope. Because of the large changes in the aircraft dynamics, a dynamic model that is stable and adequately damped in one flight condition may become unstable, or at least inadequately damped in another flight condition. For such cases, an autopilot is as an automatic control system that also provides stability augmentation.

A long range transport aircraft Boeing 777 which is equipped with a modern AFCS is shown in Figure 1.4. A few AFCS mode on this aircraft are: (1) altitude hold, (2) yaw damper, (3) approach glide slope hold, (4) Very-high-frequency Omni-directional Range (VOR) hold,

Figure 1.4: Long-range airliner Boeing 777 (https://commons.wikimedia.org/w/index.php?curid=27409879).

(5) airspeed hold, (6) vertical speed hold, (7) automatic landing, (8) automatic control of trim wheel, (9) lateral-directional SAS, and (10) automatic flight level change.

1.4.2   ATTITUDE CONTROL SYSTEMS (HOLD FUNCTIONS)

Often, aircraft lateral and directional motions are coupled. Thus, there are two basic groups of hold functions: longitudinal hold functions and lateral-directional hold functions. In the longitudinal plane, there are primarily four hold functions: pitch attitude hold; altitude hold; speed/Mach hold; and vertical speed hold. However, in lateral-directional mode, three hold functions are dominant: (1) bank angle hold or wing leveler; (2) heading angle hold; and (3) turn rate mode at level flight.

The longitudinal hold function is an efficient control method of cruising flight for a long period. Due to the burning fuel, the weight of aircraft at the beginning of the flight and at the end of flight is significantly different (usually about 20%). However, for a long endurance unmanned aircraft such as Global Hawk (Figure 1.5), the change in the aircraft weight is as much as 50%. At any weight and altitude, the lift force must be equal to aircraft weight in a straight-level flight. The Northrop Grumman RQ-4 Global Hawk is a high-altitude remotely piloted surveillance aircraft equipped with a modern AFCS.

Since the fuel is consumed during flight, the aircraft weight is constantly decreased during the flight. In order to maintain a level flight, we have to decrease the lift as well. Of the many possible solutions, only three alternatives are more practical and usually examined. In each case, two flight parameters will be held constant throughout cruise. The three options of interest for continuous decrease of the lift during cruise are: decreasing flight speed; increasing altitude; decreasing angle of attack.

Attitude control systems are often **more complex** in their operation than stability augmentation systems, since typically more control surfaces are employed.

Figure 1.5: Northrop Grumman RQ-4 Global Hawk.

1.4.3   FLIGHT PATH CONTROL SYSTEMS (NAVIGATION FUNCTIONS)

For the navigation functions, there are two groups of hold functions: longitudinal navigation functions and lateral-directional navigation functions. In the longitudinal plane, there are primarily six modes: (1) automatic flare mode; (2) approach glide slope control; (3) automatic climb and descent; (4) automatic landing; (5) terrain following; and (6) automatic flight level change. However, in lateral-directional mode, six hold functions are dominant: (1) localizer tracking; (2) VOR tracking; (3) turn coordination; (4) heading tracking; (5) tracking a series of waypoints; and (6) detect and avoid system. These navigation functions will be discussed in details in Chapter 4.

The term VOR, which stands for Very-high-frequency Omni-directional Range, is a type of radio navigation system for aircraft. The VOR station broadcasts a VHF radio signal, and data that allows the airborne receiver to derive the magnetic bearing from the station to the aircraft. This line of position is referred to as the "radial."

1.4.4   COMMAND AUGMENTATION SYSTEMS

Some autopilots improve the stability of the aircraft (SAS), while some augment the response to a control input (CAS). The slow modes (e.g., phugoid in longitudinal motion, and spiral in lateral-directional motion) are controllable by a human pilot. But since it is undesirable for a pilot to have to pay continuous attention to controlling these modes, an automatic control system is needed to provide "pilot relief." Reference [1] is presenting automatic flight control system design and provides several detailed examples.

In the category of command augmentation, three basic systems are available: command tracking; gust load alleviation system; and normal acceleration CAS. The command tracking

Table 1.1: AFCS categories

Catgory:	1	2		3		4
	Stability Augmentation Systems (SAS)	Hold Functions		Navigational Modes		Command Augmentation Systems (CAS)
Number		Longitudinal	Lateral-Directional	Longitudinal	Lateral-Directional	
1	Roll damper	Pitch Altitude hold	Bank angle hold; wing leveler	Automatic flare mode	Approach localizer hold	Pitch rate CAS
2	Pitch damper	Altitude hold	Heading angle hold	Approach glide slope tracking	VOR tracking	Lateral directional CAS
3	Yaw damper	Speed/Mach hold	Turn rate hold at level flight	Automatic flight level change	Heading tracking	Normal acceleration CAS
4	Lateral-directional SAS	Vertical speed hold		Terrain following	Turn coordination	Gust load alleviation system
5	Stall avoidance system			Automatic landing		
6	Automatic control of trim wheel			Automatic climb and descent		
7				Tracking a series of waypoints		
8				Detect and avoid system		

system are mainly divided into two modes: pitch rate tracking and roll/yaw rate tracking. Command generator tracker is known as the model following system, since it creates time varying trajectories. Table 1.1 summarizes typical common categories of AFCSs/autopilots.

In [1], the SAS and CAS are referred to as non-autopilot functions, while hold (navigation) functions are referred to as the autopilot modes. In this, text, the SAS and CAS, hold modes, and flight path control modes are all referred to as the AFCS functions. The SAS controls the aircraft motion variable when created by an atmospheric disturbance, while the CAS controls the aircraft motion variable when created by a pilot/autopilot.

1.4.5   MODE CONTROL PANEL EXAMPLES

It is beneficial to briefly review the pilot view of an AFCS in a cockpit. This will help the AFCS designer to make sure that, the pilot has an efficient experience when an AFCS mode is employed. Autopilot employs knobs, buttons, dials, displays, or other controls that allow a pilot to specify goals. To employ an AFCS, a transport aircraft pilot interacts with an aircraft using three elements: (1) Mode Control Panel; (2) Primary Flight Display (including Flight Mode Annunciator); and (3) Secondary Flight Display.

Primary flight displays often integrate all controls that allow modes to be entered for the autopilot. Knobs allow a pilot to enter modes without turning attention away from the primary flight instruments. Modes entered using the controls are transferred to the autopilot for implementation. A mode annunciator showing armed and engaged autopilot modes. The more sophisticated annunciator uses color coding to distinguish between armed and engaged autopilot functions.

Some autopilot modes cancel other modes (e.g., the altitude hold mode) when the autopilot is engaged. Every autopilot interface displays which modes are currently engaged, and most autopilot indicate an armed mode that activates when certain parameters are met, such as localizer interception. Figure 1.6 demonstrates the interface panel for a simple autopilot, which shows a selected altitude of 5,500 ft, and a few modes such as ALT (Altitude), APR (Approach), and HDG (Heading).

Figure 1.7 exhibits the Boeing 777 Mode Control Panel which includes 4 select windows (1. IAS/Mach, 2. Heading Track, 3. V/S, FPA, and 4. Altitude), 12 buttons (including Autopilot), and other switches. Figure 1.8 illustrates primary flight display of a Boeing 777 where thee modes of SPD (Speed hold), LOC (Localizer hold), and V/S (Vertical speed hold) are engaged. For instance, in Boeing 747, Command modes can be armed or engaged when an autopilot (A/P) is engaged in command and/or one or both flight director (F/D) switches are ON. The Take-off mode is a F/D only mode, while the automatic flare during the approach (APP) mode is a dual A/P only maneuver.

The interface panel for a Garmin GFC 700 is shown in Figure 1.9. This fully digital dual-channel flight control system features capabilities such as cruise speed control, flight level change, yaw damper, and automatic electric trim. The GFC 700 mode selector provides dedicated knobs

Yaw Damper   Heading Navigation Approach   Altitude   Altitude Selection
        Autopilot   Radio   Knob

Figure 1.6: Interface panel for a simple autopilot.

Figure 1.7: Boeing 777 mode control panel.

for heading and altitude selection and buttons for mode changes. This AFCS has been used in several aircraft including GA aircraft Cessna 350 and business jet Gulfstream G700.

1.5   FLYING QUALITIES

1.5.1   FUNDAMENTALS

In the design of an AFCS, the designer must consider the flying qualities specifications and criteria. Using a right controller/compensator, the closed-loop transient response of an aircraft should be made to have dominant modes that are known to provide acceptable handling qualities for a human operator. In lateral-directional motions, the Dutch roll mode should be fast and adequately damped so that the aircraft will quickly reorient itself after a directional disturbance. Both longitudinal modes—phugoid and short period—should be stabilized.

The presentation in this section is intended to give only the general flavor of the flying qualities. Typical guidance used in developing aircraft platform system specifications for flying qualities are MIL-F-8785, MIL-HDBK-1797, MILHDBK-516, and ADS-33-PRF. Many of

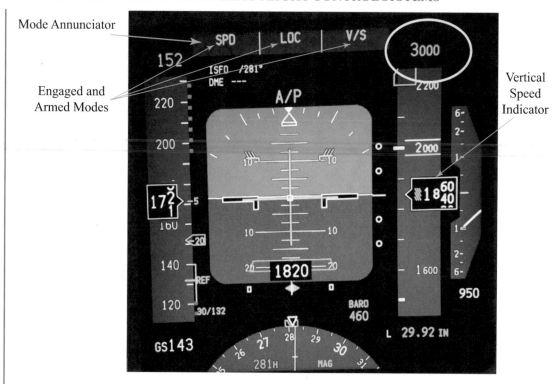

Figure 1.8: Primary flight display of a Boeing 777.

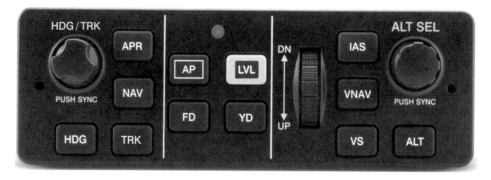

Figure 1.9: Interface panel for a Garmin GFC 700.

the flying or handling qualities specification for manned aircraft (e.g., stick force) do not directly apply to UAVs and autopilot design. However, for the case of a UAV, which is remotely controlled by an operator in the ground station, there are a number of requirements which influence the stability and controllability of the UAV. Moreover, to allow for tradeoffs between the air vehicle and sensor/payload subsystems, flying qualities specifications have to provide flexibility and guidance in tailoring air vehicle requirements.

Moreover, special consideration must be given to the way in which the autopilot is engaged and disengaged, so that dangerous transient motions are not generated. Another area of interest for handling qualities in the design of autopilot is the process through which the autopilot mode is switched from one to another. The differences between manned aircraft and UAV flying qualities mainly relate to primary and secondary flight control systems, data link time delays, system failure states, and flight display requirements. The flying qualities of an airplane are strongly connected to its control surfaces, mainly in size of area, arm, and defection angle. Compliance with the requirements can be demonstrated through analysis, simulation, and flight tests.

1.5.2   CLASSES, CATEGORIES, AND ACCEPTABILITY LEVELS

The top-level requirement in MIL-F-8785C and MIL-STD-1797B is for aircraft to have Level 1 flying qualities in the Operational Envelope, and Level 2 flying qualities in the Service Envelope. The detailed criteria in the lower-level requirements define Level 1, 2, or 3 flying qualities. Detailed performance to meet Level 1, 2, or 3 flying qualities generally varies as a function of aircraft class and flight phase.

For manned aircraft, there are four classes, which are based on size, maneuverability, and missions. The flight envelope usually covers altitude, airspeed, and normal acceleration. Class I is for small and light aircraft; Class II is for medium-weight, low to medium maneuverability aircraft; Class III are large and heavy airplanes such as large transport aircraft; and Class IV is for high maneuverability aircraft such as fighters.

However, for UAVs, these should be replaced with: (I) micro UAVs; (II) mini UAVs; (III) small UAV; (IV) medium weight UAVs; (V) heavy weight UAVs; and (VI) UCAVs. These are for fixed-wing UAVs; a new category must also be defined for quadcopters (i.e., class VII).

Flight phases are divided into three Categories: A, B, and C. Category A includes flight phases that require precision/rapid maneuvering such as combat, or terrain following. Category B includes flight phases that are not terminal and do not require precision maneuvering such as cruise and loiter. Category C contains the terminal flight phases, such as take-off and landing. These categories could be kept the same for UAVs. Without humans onboard, higher loss rates than piloted aircraft is acceptable.

A new set of specifications should be prepared for the flight modes: (a) remotely controlled through line-of-sight; (b) remotely controlled through display; (c) single-mode autopilot (e.g., altitude-hold mode); and (d) fully autonomous. Each of these flight modes should have unique flying qualities specifications, due to the level of autonomy.

1.6   QUESTIONS

1.1.   Name primary subsystems of an autopilot.

1.2. What are the main functions of an autopilot?

1.3. What are the main four functions for a closed-loop (i.e., negative feedback) control system?

1.4. What are the two primary goals of an AFCS?

1.5. Briefly describe the relations between a pilot and an AFCS in an aircraft.

1.6. Is it possible to have an active mode of an AFCS, when the pilot is manually controlling an aircraft? If yes, provide one example mode.

1.7. Briefly describe an irreversible control system.

1.8. What is the primary function of the guidance system?

1.9. What is the primary function of the navigation system?

1.10. List phases of a typical full flight.

1.11. Define autonomy.

1.12. List 11 levels of autonomy.

1.13. Define Laplace transform.

1.14. Define transfer function.

1.15. Provide a typical proper second order transfer function.

1.16. What does SISO stand for?

1.17. What does MIMO stand for?

1.18. Provide the standard form for a pure second-order dynamic system.

1.19. What are the poles and zeros of a transfer function?

1.20. What is dynamic modeling?

1.21. What are two methods for simplifying a nonlinear coupled dynamic model?

1.22. Describe the state space model.

1.23. What are the main aerodynamic forces?

1.24. What are the main aerodynamic moments?

1.25. List various types of dynamic models of an aircraft.

1.26. Write the linear dynamic model (state-space) for longitudinal mode for a fixed-wing aircraft.

1.27. Write the linear dynamic model (state-space) for lateral-directional mode for a fixed-wing aircraft.

1.28. What does the aircraft dynamic model consist of?

1.29. Draw an aircraft, and show all aerodynamic forces and moments in this figure.

1.30. What are the outputs for decoupling longitudinal motion from lateral-direction motions?

1.31. Briefly describe the features of a single axis autopilot.

1.32. List important modes of the AFCS in the transport aircraft MD-11.

1.33. Name five categories/functions for AFCSs.

1.34. Name at least three stability augmentation systems (SAS) in an autopilot.

1.35. Name at least four longitudinal hold functions in autopilots.

1.36. Name at least three lateral-directional hold functions in autopilots.

1.37. Name at least four longitudinal navigation functions in autopilots.

1.38. Name at least three command augmentation systems (CAS) in autopilots.

1.39. List at least five modes of AFCS for a transport aircraft Boeing 777.

1.40. To employ an AFCS, a transport aircraft pilot interacts with an aircraft using three elements. Name them.

1.41. Provide at least two references for flying qualities.

1.42. List manned aircraft classes with regard to flying qualities.

1.43. List UAV classes with regard to flying qualities.

1.44. List flight phases categories with regard to flying qualities.

CHAPTER 2

Closed-Loop Control Systems

2.1   INTRODUCTION

One of the main subsystems within an AFCS is the control system. In general, a closed-loop (i.e., negative feedback) control system tends to provide four functions: regulating; tracking; stabilizing; and improving the plant response. The control system is used to keep an aircraft on a predetermined course or heading, necessary for the mission. The control system is using the vehicle state information provided by the on-board sensors to drive the control surface actuators (i.e., servos). The control system should control the direction of the motion of the vehicle or the orientation of the velocity vector.

Despite poor and gusty weather conditions, an aircraft must maintain a specified heading and altitude in order to reach its destination safely. In addition, in spite of rough air, the trip must be made as smooth as possible for a low load factor. The problem is considerably complicated by the fact that the UAV has six degrees of freedom. This fact makes control more difficult than the control of a ship, whose motion is limited to the surface of the water.

An essential activity in the detail design is the development of a functional description of the control system components to serve as a basis for identification of the resource necessary for the system to accomplish its mission. Such a function may ultimately be accomplished through the use of equipment, software, facilities, data, or various combinations thereof. The control system design must meet the controllability requirements prescribed for maneuvers as desired by the customer or standards.

With advances in computer technology, and the introduction of new mathematics theories in nonlinear systems, more applications of advanced control system design techniques such as robust and nonlinear control are seen in the literature. In this chapter, the fundamentals of closed-loop control system, flight control requirements, and a few popular control laws are presented.

2.2   FUNDAMENTALS OF CONTROL SYSTEMS

The automatic flight control system must fundamentally be of a closed-loop form, which employs negative feedback(s). In this section, fundamentals of control systems, control concepts, basic definitions, major elements, and general design techniques are presented.

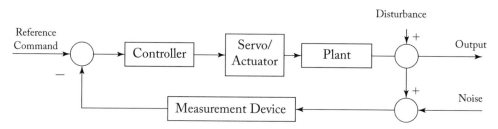

Figure 2.1: Block diagram of a closed-loop control system including disturbance and noise.

2.2.1   DEFINITIONS AND ELEMENTS

The essential feature of an automatic flight control system is the existence of feedback loop to give a good performance. This is a closed-loop system (i.e., feedback control system); if the measured output is not compared with the input, the loop is open. Usually, it is required to apply a specific input to a system and for some other part of the system to respond in the desired way. The error between the actual response and the ideal response is detected and fed back to the input to modify it so that the error is reduced.

In terms of the AFCS application, a flight controller in a feedback control system is used to control the aircraft motion. Two typical signals to the system are the reference flight path, which is set by the autopilot, and the level position (attitude) of the airplane. The ultimately controlled variable is the actual course and position of the aircraft. The output of the control system, the controlled variable, is the aircraft heading.

Figure 2.1 illustrates a basic Single-Input-Single-Output (SISO) closed-loop system. The simplest linear closed loop system incorporates a negative feedback, and will have one input and one output variable. A closed-loop system is generally made up of four basic elements: plant; controller; actuator (or servo); and measurement device (or sensor). In general, both input and output vary with time, and the control system can be mechanical, pneumatic, hydraulic, and electrical in operation, or any combination of these or other power sources.

It is well known that the atmosphere is a dynamic system that produces lots of disturbances throughout the aircraft's flight. Disturbance is the unwanted signal (e.g., gust) that tends to affect the controlled variable. The disturbance may be introduced into the system at many places.

It is also important to note that all measurement devices (including gyros and accelerometers) have some kinds of noise that must be filtered. Noise is the unwanted signal (e.g., engine vibration) that tends to affect the measured variable. A filter may be utilized to get rid of noise in the measurement device.

Noise, by definition, is any undesired signal within the system, whether natural or man-made. The principal components of system noise are wide-band noise, 1/f noise, and interference. Wide-band noise consists of thermal noise, shot noise, and partition noise. Interference noise is man-made noise that can be reduced by proper circuit layout, shielding, and grounding techniques. The 1/f noise component, which exists in all natural processes, shows the accumulated effects of small changes. Different measurement devices have different noises from a variety of sources.

In general, noise can be removed from an accelerometer via a low-pass filter. The noise spectrum is white noise, meaning that all frequencies are present equally, at least throughout the radio spectrum. To estimate a signal of interest, or the state of a system in the presence of additive noise, a filter is used. In implementing control law, there are two main approaches: analog control and digital control.

The real aircraft behavior is nonlinear and has uncertainties. An aircraft is basically a nonlinear complex system. Furthermore, its dynamics and equations of motion are also nonlinear. In this section, some common nonlinear systems phenomena are presented. Later on, they are applied to the design of the UAV autopilot. Systems containing at least one nonlinear component are called nonlinear systems. Basically, there are two types of nonlinearities: continuous and discontinuous (hard). Hard nonlinearities include: coulomb friction; saturation; dead-zone; backlash; and hysteresis. In another classification, nonlinearities could be inherent (natural) or intentional (artificial).

It is also important to note that all measurement devices (including gyros and accelerometers) have some kinds of noise that must be filtered. It is well known that the atmosphere is a dynamic system that produces lots of disturbances throughout the aircraft's flight. Finally, since fuel is expensive and limited, and actuators have dynamic limitations, optimization is necessary in control system design. Therefore, it turns out that only a few design techniques, such as robust nonlinear control, are able to satisfy all safety, cost, and performance requirements. However, due to cost, and complexity, many UAV designers adopt a more conventional control architecture. Two conventional controller design tools/techniques are: root locus technique and frequency domain techniques. The interested reader may refer to Refs. [38] and [39] for more details.

2.2.2   CONTROL LAWS

The controller is designed based on a control law. Some typical control laws are: linear; nonlinear; optimal; adaptive; and robust. A summary of the dynamic systems and the different control system design techniques is shown in Figure 2.2. In implementing control law, there are two main approaches: analog control and digital control.

The design of control system by the root locus method is determining the controller gain value such that all poles are within the target area, and all design requirements are met. The target area is constructed by applying all design requirements such as rise time, settling time, maximum overshoot, natural frequency, and damping ratio. The addition of a pole to the open loop TF has the effect of pulling the root locus to the right, tending to lower the system's relative stability and to slow down the settling of the response. The zero has the opposite effect.

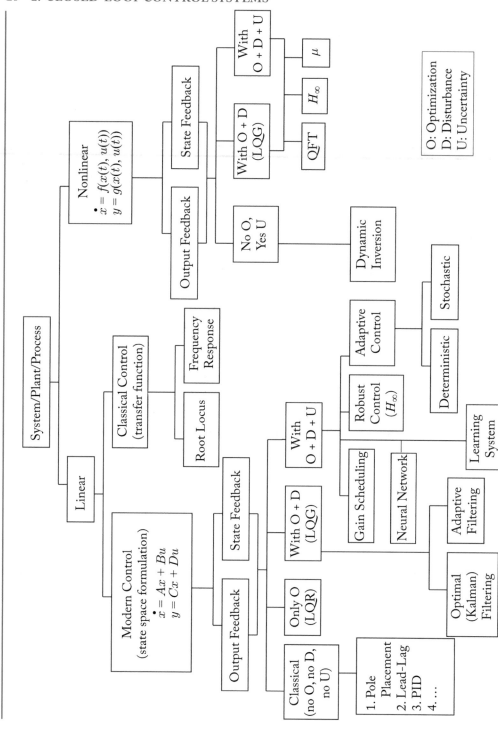

Figure 2.2: Control system design techniques.

In practice, this technique may indicate that the desired performance cannot be achieved just by a simple gain controller. Moreover, in some cases, we may find out that the system will not be stable for all values of the gain. This implies that a simple gain is not a solution, and a more advanced controller should be designed. A series of calculations are involved in determining poles and zeros of the compensator (e.g., lead compensation). One of the primary goals in the root locus technique is to have the dominant closed loop poles at the desired location in the s plane, so that the performance specifications are met. The Matlab has the command "rlocus" to construct the root locus. Moreover, the Matlab command "sisotool" is creating the root locus as well as the time response of the system.

2.2.3   CONTROLLER CONFIGURATIONS AND CONTROL ARCHITECTURES

In general, there are six basic controller configurations in control system compensations: (1) series or cascade compensation (Figure 2.1); (2) feedback compensation; state-feedback control; (4) series-feedback compensation (two-degree of freedom (2DOF)); (5) forward compensation with series compensation (2DOF); and (6) forward compensation (2DOF). Each configuration is best for a particular circumstance, and has unique features including cost, complexity, performance, and effectiveness.

For instance, one application for the two-degree of freedom controller is the *time-scale separation*; where the flight (state) variables are divided into two groups: a. slow-states, and b. fast states. In this technique, the dynamics to be controlled are separated into fast and slow variables. In application to flight vehicle dynamics, the control technique (e.g., LQR) modifies the problem into a two time-scale problem. The two time scales are estimated by assuming that there is a significant frequency separation between the fast and slow states. The number of slow and fast states depends on the number of controls. If there are three control surfaces, roll, yaw, and pitch rates (p, q, r) are defined as fast dynamic variables, which are controlled by three inputs: aileron, elevator, and rudder.

Angles of attack, sideslip angle, and bank angle (α, β, ϕ) are defined as slow dynamic variables. The speed, pitch angle, and heading angle are defined as very slow dynamic variables. The assumption is that the very slow states are not time-varying when compared with the fast and slow dynamic variables.

2.2.4   FLIGHT CONTROL MODES

Two primary prerequisites for a safe flight are stability and controllability. The control system is not only able to control the UAV, but also is sometimes expected to provide/augment stability. Flight stability is defined as the inherent tendency of an aircraft to oppose any input and return to original trim condition if disturbed. When the summation of all forces along each three axes, and summation of all moments about each three axes are zero, an aircraft is said to be in trim or equilibrium. In this case, aircraft will have a constant linear speed and/or a constant angular speed.

Control is the process to change the aircraft flight condition from an initial trim point to a final or new trim point. This is performed mainly by autopilot through moving the control surfaces/engine throttle. The desired change is basically expressed with a reference to the time that takes to move from initial trim point to the final trim point (e.g., pitch rate; q, and roll rate, p).

An aircraft is capable of performing various maneuvers and motions; they may be broadly classified into three main groups: longitudinal control; lateral control; and directional control. In the majority of aircraft, longitudinal control does not influence the lateral and directional control. However, lateral and directional control are often coupled; any lateral motion will often induce a directional motion; and; any directional motion will often induce a lateral motion. The definition of these motions is as follows.

1. *Longitudinal control*: Any rotational motion control in the x-z plane is called longitudinal control (e.g., pitch about y-axis, plunging, climbing, cruising, pulling up, and descending). Any change in lift, drag and pitching moment have the major influence on this motion. The pitch control is assumed as a longitudinal control. Two major longitudinal control inputs are the elevator (δ_E) and engine throttle (δ_T).

2. *Lateral control*: The rotational motion control about x-axis is called lateral control (e.g., roll about x-axis). Any change in wing lift distribution and rolling moment have the major influence on this motion. The rolling control is assumed as a lateral control. The primary lateral control input is the aileron deflection (δ_A). However, the rudder (δ_R) deflection has an indirect influence on this motion too.

3. *Directional control*: The rotational motion control about z-axis and any motion along y-axis is called directional control (e.g., yaw about z-axis, side-slipping, and skidding). Any change in side-force and yawing moment have the major influence on this control. The yaw control is assumed as a directional control. A level turn is a combination of lateral and directional motions. The primary directional control input is the rudder deflection (δ_R). However, the aileron deflection (δ_A) has an indirect influence on this motion too.

In a conventional aircraft, three primary control surfaces are used to control the physical three-dimensional attitude of the aircraft, the elevators, rudder, and ailerons. In a conventional aircraft, the longitudinal control (in the x-z plane) is performed through a longitudinal control surface or elevator. The directional control (in the x-y plane) is performed through a directional control surface or rudder. The lateral motion (rolling motion) is executed using aileron. Hence, there is a direct relationship between the control system and control surfaces. A graphical illustration of these three control motions is sketched in Figure 2.3. An aircraft has six degrees of freedom (i.e., three linear motions along x, y, and z; and three angular motions about x, y, and

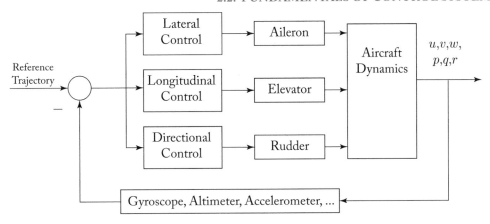

Figure 2.3: Flight control system with conventional control surfaces.

z). Thus, there are normally six outputs in an actual flight; six examples are three linear velocities (u, v, w) and three angular rates (p, q, r).

The primary idea behind the design of flight control surfaces is to position the surfaces; so that they function primarily as moment generators. They provide three types of rotational motion (roll, pitch, and yaw). Variations to this classical configuration lead to some variations in the arrangements of these control surfaces.

2.2.5   SENSORS

There are various measurement devices to measure the flight variables such as airspeed, pitch angle, heading angle, bank angle, linear accelerations (normal, lateral, and longitudinal), angular rates (pitch, roll, and yaw rates), altitude, and position. The measured flight data are recorded, and may be stored in a data storage element, which may be conveniently read by the user (in real-time, or off-line). To utilize airframe output quantities (u, v, or β, w or α, p, q, r and their time derivatives) for automatic control, they must be sensed, or "picked up," by some means. This section presents a discussion of some of the devices, which are used for this purpose.

If an interface with telemetry data systems is available, a user in the ground station may access the data in real time. Typical measurement devices (sensors) are: (1) attitude gyroscope; (2) rate gyroscope; (3) pitot-tube; (4) altimeter; (5) magnetometer; (6) compass; (7) accelerometer; (8) GPS; (9) flow incidence angle sensors, and (10) airspeed sensors.

Several flight instruments employ the properties of gyroscopes for their operations. The most common instruments containing gyroscopes are the turn coordinator, heading angle indicator, the attitude indicator, and angular velocity indicator. There are two fundamental properties of gyroscopic action: rigidity in space and precession. Rigidity in space refers to the principle that a gyroscope remains in a fixed position (memory) in the plane in which it is spinning. Precession is the tilting or turning of a gyro in response to an applied force. The reaction to this force does not occur at the point at which it was applied; rather, it happened at 90° to the direction of rotation. This principle allows the gyro to determine a rate of turn by sensing the amount of force created by a change in direction.

A directional gyro is a heading indicator, while a vertical gyro is a pitch indicator. Moreover, a vertical gyro is used for measuring bank angle, while a directional gyro is used for the heading reference. The directional gyroscope is used as the error-measuring device. Employing two gyros allows a designer to provide control of both heading and attitude (level position) of the airplane. A time constant of a few seconds represents a lag in the response of the rate gyro. Typical sensitivity of a rate gyro is about 1 volt/(deg/sec).

The error that appears in the gyro as an angular displacement between the rotor and case is translated into a voltage by various methods, including the use of transducers such as potentiometers. Augmented stability for the aircraft may be desired in the control system by rate feedback. In other words, in addition to the primary feedback, which is the position of the airplane, another signal proportional to the angular rate of rotation of the airplane around the vertical axis is fed back in order to achieve a stable response. A rate gyro is used to supply this signal.

External probes (sensors) are employed to sense air data parameters such as Mach number and dynamic pressure. The air data must be derived from the navigation system and a stored mathematical model of the atmosphere.

In general, the accelerometer is less noisy and more reliable than an angle of attack sensor in controlling longitudinal motions. An accelerometer is an internal (within the fuselage) sensor, with higher reliability and lower noise than the external alpha sensor.

The traditional gyros are made of spinning wheels, but progress in microelectromechanical systems (MEMS) resulted in new effective sensors. MEMS are a new group of low-cost, lightweight sensors for variety of applications from measurement of pressure/temperature to acceleration/attitude. These sensors combine the benefits of high-precision in electronic circuitry and high-load capability of mechanical systems in micro-size level. For instance, the MEMS gyroscopes (new micro-gyros that do not have spinning wheels) measure changes in the forces acting on two identical masses that are oscillating and moving in opposite directions. Moreover, A MEMS magnetic field sensor is a small-scale MEMS sensor for detecting and measuring magnetic fields. The Garmin GFC 700 **MEMS**-based automatic flight control system is employing Attitude and Heading Reference System (AHRS) data.

2.3 SERVO/ACTUATOR

2.3.1 TERMINOLOGY

Every autopilot system features a collection of electromechanical devices, called servos/actuator, that actuate the aircraft control surfaces. Another important element in the aircraft flight control is the actuator which deflects (pushes and pulls) the control surfaces (e.g., elevator). There are a few respective terms used by various groups/disciplines, which is briefly reviewed here. A servo is a rotary/linear actuator that allows for precise control of angular/linear position, velocity, and acceleration. Electric motors as servos translate electrical commands into motion. The term servo is mainly found in the electrical engineering terminology (also servomotor), as well as the radio/remotely controlled airplane modelers' community. Optical encoders are usually used in servo systems to detect the position and control the movement of power drives and sensors such as radar.

The term servomotor is used when an electric motor is coupled with a sensor for position feedback. A servomechanism is a feedback control system in which the controlled variable is a (mechanical) position, velocity, torque, frequency, etc. It consists of an electric motor, mechanical linkage, and a microcontroller. An actuator is a mechanical/hydraulic device used as a mechanism to translate mechanical motion (often rotary) into linear motion. The term "actuator" is mainly found in the aerospace engineering terminology.

When a signal is received from the microcontroller/autopilot to deflect a control surface, the actuator will implement this command. The source of actuator power is either an electric, mechanical, hydraulic, or pneumatics device. For mini to medium-small UAVs (and also quadcopters), an electric motor is frequently used. However, for a large and heavy aircraft and UAV (e.g., Global Hawk), a hydraulic actuator is more effective. For the case of hydraulic/pneumatic power, the servo is a hydraulic/pneumatic cylinders (or valve); not an electric motor. A comparison between a number of features for hydraulic actuators and servomotors are provided in Table 2.1.

Small servos are of two types: analog (frequency of 50 Hz) and digital (frequency of 100 Hz). Digital servo looks the same as analog servo (motor, gear, and potentiometer), but has a microprocessor. Digital servo is much faster than the analog servo, but analog servo has overshoot. Digital servo is more expensive, more accurate, and are faster, thus, it consumes more electric energy. Small servos are available for various output torques (so, various sizes and weights).

2.3.2   ELECTRIC MOTOR

Another electromechanical device that are used to deflect aircraft control surfaces (as actuators) is electric motor. An electric motor is an electromechanical device that uses electric energy to produce mechanical rotational motion. It may be powered by: direct current DC (e.g., a battery), or alternating current (AC). Electric motors may be classified by: the source of electric power; their internal construction; and their application. Two main parts of an electric motor is rotor coil and magnet. A well-designed motor can convert over 90% of its input energy into useful power.

A *servomotor* consists of an electric motor coupled with a sensor to measure the position/speed to create a feedback, (and may have a microprocessor/microcontroller). It also requires a controller and measurement device (e.g., encoder or potentiometer) to create a closed-loop control system. The input to its control is a signal (either analog or digital) representing the position commanded for the output shaft. The very simplest servomotors use position-only

Table 2.1: Features of two types of servos

No	Actuator Type	Electric Engine	Hydraulic Actuator
1	Power	Electric	Hydraulic
2	Motion	Rotation	Linear
3	Typical physical form		
4	Output	Torque	Force
5	Capability	Torque range: 0.1–100 N.m	Force range: 10–1,000 N
6	Overall weight	Lighter	Heavier
7	Medium	Wire	Tube/pipe
8	Signal	Electricity	Oil
9	Maintenance	More labor-intensive	Less labor-intensive

sensing via a potentiometer and bang-bang control of the motor. More accurate servomotors use optical rotary encoders to measure the angular speed of the output shaft to control the motor speed.

In general, there are four types of electric motors: (1) brushless DC (direct current) motors; (2) brushed DC motors; (3) brushless AC (alternating current) motors; and (4) induction AC motors. The simplest one is brushed DC motor which has a permanent magnet and is of low cost. The brushless DC motor is synchronous, and is a good candidate for small industrial application and is typically electronically commutated. It uses electronic commutation system rather than a mechanical commutator and brushes. In brushless DC motors, the current to torque, and voltage to rpm have linear relationships. The AC motors are used in large heavy aircraft, while DC motors are used in micro to small aircraft, as well as quadcopters.

A servo-motor, also referred to as smart motor, is combination of an electric motor, an electromechanical sensor, an encoder and a controller as one package. Servomotors are generally used as a high-performance alternative to the stepper motors (or step motors) which have some inherent ability to control position. A main difference between servomotor and stepper motor is the feedback, so servomotors have a closed-loop control system (due to encoder and controller), while step motors have an open-loop system.

2.3.3   HYDRAULIC ACTUATOR

Another type of actuator, which is employed for the deflection of control surfaces, are hydraulic actuators (i.e., jack). The hydraulic actuator is a cylinder that converts hydraulic power into useful mechanical (often linear) motion. Such an actuator uses hydraulic pressure as a mechanical force to move objects like controls. It consists of two basic mechanisms: a control device (variable throttles, gates or valves) and an actuating device (e.g., the piston of an actuating mechanism). In a hydraulic actuator, the working fluid (i.e., oil) from a pressure line enters the control system through the constant throttles and is carried to the variable throttles and to the actuator chambers.

The electrical input signal—when passed through an electromechanical converter—controls the position of the gate slide. The displacement of this slide changes the cross-sectional ratio of the actuating openings through which the pressure fluid passes. Simultaneously, the pressures in the actuator chambers are changed, thus causing a displacement of the slide valves.

The hydraulic actuators compared with other types of power amplifiers (e.g., electromechanical or pneumatic amplifiers), offer the advantage of a low metal weight per unit of power, often not exceeding 50 g per kilowatt of output power. The power amplification of a hydraulic actuator is very high (about 100,000). Hydraulic actuators with a load feedback substantially improve the dynamic features, and the efficiency of hydraulic control systems.

Other candidates for linear actuators are electromechanical and pneumatic ones. Both of these actuators have advantages and disadvantages compared with hydraulic actuators. The reliability of hydraulic actuators is higher than that of pneumatic ones, which have a higher cost.

2.3.4   DELAY

To represent the actuator's delay (i.e., lag) due to a physical limitation, we may use a transfer function or state-space model. A regular aircraft actuator for control surfaces (e.g., aileron servo) is modeled with a first-order transfer function (lag), G_A, which is of the form:

$$G_A(s) = \frac{K}{s + K},$$ (2.1)

where K represents the inverse of the time constant (τ) of the actuator:

$$\tau = \frac{1}{K}.$$ (2.2)

A typical value for the time constant of an aircraft control surface actuator is about 0.02–0.1 sec, so the K frequently varies between 10 and 50 ($10 < K < 50$). Recall that the time constant is defined as the time that takes (i.e., delays) the response of an element to reach 63% of the steady-state value. The smaller the time constant, the faster (more desired) the actuator. A first-order model may also represent the dynamic behavior of an electric motor (used in most quadcopters). In frequency domain, the constant K represents the break frequency. The higher the break frequency, the greater the cost of the actuator.

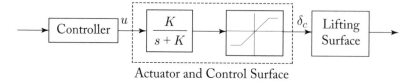

Figure 2.4: Scheme of the control surface actuator.

In terms of state-space representation, the transfer function (Equation (2.1)) which represents the delay, can be converted to the following form:

$$\dot{x} = -Kx + Ku$$
$$y = x. \tag{2.3}$$

Either of the models may be utilized in the control system design and simulation.

2.3.5   SATURATION

All control surfaces have a limit for maximum deflection (about $\pm 30°$). To prevent a control surface (e.g., aileron) not to exceed the desired limit, a physical stop (*limiter*) is arranged. For instance, the stops for the ailerons are just a piece of aluminum bar riveted to the aileron hinge bracket. This hard stop is modeled in Matlab/Simulink with a saturation block. Hence, a control surface and its actuator are modeled (Figure 2.4) with a first-order system (Equation (2.1)) plus a limiter. Thus, we need to bound the signal entering the actuator.

2.4   FLIGHT CONTROL REQUIREMENTS

There are primarily three flight control requirements: (1) longitudinal control requirements; (2) lateral/roll control requirements, and (3) directional control requirements. In this section, these requirements are presented. These requirements must be met by the flight control system. The description and analysis of aircraft modes show that Automatic Flight Control Systems (AFCSs) can be divided into different categories. One category includes modes that involve mainly the rotational degrees of freedom, whereas some categories involve the translational degrees of freedom.

2.4.1   LONGITUDINAL CONTROL REQUIREMENTS

An aircraft must be longitudinally controllable, as well as maneuverable within the flight envelope. In a conventional aircraft, the longitudinal control is primarily applied through the deflection of elevator (δ_E) and engine throttle setting (δ_T). There are two groups of requirements in the aircraft longitudinal controllability: 1. required actuator force (in linear actuator); or torque (in servomotor), and 2. aircraft response to the control surface deflection input. In order to de-

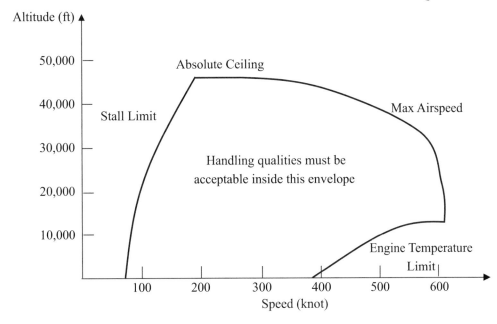

Figure 2.5: A typical operational flight envelope.

flect the elevator, the pilot must apply a force to stick/yoke/wheel and hold it (in the case of an aircraft with a stick-fixed control system). In an aircraft with a stick-free control system, the pilot force is amplified through such devices as tab and spring.

The aircraft response in the longitudinal control is frequently expressed in terms of pitch rate (q). However, the forward speed and angle of attack would be varied as well. The most critical flight condition for pitch control is when the aircraft is flying at a low speed. Two flight operations, which feature a very low speed, are take-off and landing. Take-off control is much harder than the landing control due to the safety considerations. A take-off operation is usually divided into three sections: (1) ground section; (2) rotation or transition; and (3) climb. The longitudinal control in a take-off is mainly applied during the rotation section, which the nose is pitched up by rotating the aircraft about main gear.

The control surfaces must be designed such that aircraft possesses acceptable flying qualities anywhere inside the operational flight envelope; and allowable cg range, and allowable aircraft weight. The operational flight envelopes define the boundaries in terms of speed, altitude, and load factor within which the aircraft must be capable of operating in order to accomplish the desired mission. A typical operational flight envelope for a large aircraft is shown in Figure 2.5.

2.4.2 ROLL CONTROL REQUIREMENTS

Roll or lateral control requirements govern the aircraft response to the aileron deflection; thus, the requirements are employed in the design of aileron. It is customary to specify roll power in terms of the change of bank angle achieved in a given time, in response to a step function in roll command. Thus, the aircraft must exhibit a minimum bank angle within a certain specified time in response to aileron deflection. The required bank angles and time are specified in tables for various aircraft classes and different flight phase (see Refs. [9] and [10]).

Roll performance in terms of a bank angle change ($\Delta\phi$) in a given time (t) is specified in tables (Ref. [9]) for Class I–IV aircraft. The notation "60 degrees in 1.3 seconds" indicates the maximum time it should take from an initial bank angle (say 0°) to reach a bank angle which is 60° different than the initial one, following the full deflection of aileron. It may also be interpreted as the maximum time it should take from a bank angle of $-30°-+30°$. For Class IV aircraft, for level 1, the yaw control should be free. For other aircraft and levels, it is permissible to use the yaw control to reduce any sideslip, which tends to retard roll rate. Such yaw control is not permitted to induce sideslip, which enhances the roll rate.

2.4.3 DIRECTIONAL CONTROL REQUIREMENTS

In a conventional aircraft, directional control is usually maintained by the use of aerodynamic controls (e.g., rudder) alone at all airspeeds. There are a number of cases that directional control must be achievable within a specified limits and constraints. Directional control characteristics shall enable the pilot to balance yawing moments, and control a yaw and a sideslip. Sensitivity to yaw control pedal forces shall be sufficiently high that directional control and force requirements can be met and satisfactory coordination can be achieved without unduly high pedal forces, yet sufficiently low, that occasional improperly coordinated control inputs will not seriously degrade the flying qualities.

In a multi-engine aircraft, at all speeds above 1.4 Vs, with asymmetric loss of thrust from the most critical factor while the other engine(s) develop normal rated thrust, the airplane with yaw control pedals free may be balanced directionally in steady straight flight. The trim settings shall be those required for wings-level straight flight prior to the failure. When an aircraft is in directional trim with symmetric power/thrust, the trim change of propeller-driven airplanes with speed shall be such that wings-level straight flight can be maintained over a speed range of ±30 percent of the trim speed or ±100 knots equivalent airspeed, whichever is less (except where limited by boundaries of the Service Flight Envelope) with yaw-control-device (i.e., rudder). In the case of one-engine-inoperative (asymmetric thrust), it shall be possible to maintain a straight flight path throughout the Operational Flight Envelope with yaw-control device (e.g., rudder) not greater than actuator's maximum force, without re-trimming.

2.5   CONTROL LAWS

Due to the nonlinearities and uncertainties of the aerial vehicle dynamics, various advanced control techniques, such as neural network, fuzzy logic, sliding mode control, robust control, and learning systems, have been used in autopilot systems to guarantee a smooth desirable flight mission. In this section, five control laws are briefly introduced.

2.5.1   PID CONTROLLER

One form of controller widely used in industrial process control is called a three-term, or PID (Proportional-Integral-Derivative) controller. In this controller, three operations are applied on the error signal: (1) proportionally (P) amplified; (2) integrated (I); and (3) differentiated (D). Thus, the control signal, $u(t)$, in time domain is:

$$u(t) = K_P(e(t)) + K_I \int e(t)dt + K_D \frac{de(t)}{dt}. \tag{2.4}$$

Hence, the controller has three terms: proportional, integral, and derivative. In s-domain, this controller has a transfer function:

$$G_c(s) = K_P + \frac{K_I}{s} + K_D s. \tag{2.5}$$

Various performance deficiencies may be corrected by employing the right values of PID gains. This type of controller is effective, low cost, and easy to apply. Thus, it is even used in aircraft autopilot. References [38] and [39] provide a technique to determine PID gains.

2.5.2   OPTIMAL CONTROL – LINEAR QUADRATIC REGULATOR

The optimal control [40] is based on the optimization of some specific performance criterion, or Performance Index, J. In this technique, no disturbance, noise, or uncertainty is considered. The performance of a control system, written in terms of the state variables, can be expressed as:

$$J = \int_0^{t_f} g(x,u,t)dt. \tag{2.6}$$

We are interested in minimizing the error of the system; any deviation from equilibrium point is considered as an error. For this goal, an error-squared performance index is defined—in order to include any positive or negative deviations. For a system with one state variable, x_1, we have

$$J = \int_0^{t_f} [x_1(t)]^2 \, dt, \tag{2.7}$$

an optimization technique for a dynamic system in state-space format is defined. The Linear Quadratic Regulator (LQR) is an optimal controller. The LQR problem is simply defined as follows. The system of interest is of the form:

$$\dot{x} = Ax + Bu$$

$$y = Cx + Du \quad , \quad x(0) = x_o. \tag{2.8}$$

Given the *weighting matrices* Q and R, the design task is to find the optimal control signal $u(t)$ such that the quadratic cost function,

$$J = \frac{1}{2} \int_0^\infty \left(x^T Q x + u^T R u \right) dt \tag{2.9}$$

is minimized. The solution to this problem is:

$$u = -Kx \tag{2.10}$$

where

$$K = R^{-1} B^T P \tag{2.11}$$

and P is the unique, positive semi-definite solution to the *Algebraic Riccati Equation* (ARE):

$$PA + A^T P + Q - PBR^{-1}B^T P = 0. \tag{2.12}$$

The matrices Q and R are the weight for state and input variables, respectively. They are determined based on the cost function. The engineering judgment skill must be utilized in the selection of Q and R. Tuning techniques are recommended in the determination of design parameters. For instance, Q and R must be such that the detectability and observability requirements are met. The linear-quadratic regulator (LQR) is a popular optimal control technique that has been successfully applied to control several aircraft configurations.

2.5.3   GAIN SCHEDULING

An effective control law for a time-varying dynamic model is gain scheduling, at which the gains should be scheduled as a function of time. A few cases for an aircraft are: (1) when the fuel is rapidly burned, both vehicle weight and its center of gravity are varying with time; (2) when the flight Mach number is considerably changing, the aerodynamic model is varying with time; (3) when the flight altitude is changing, the engine power/thrust are significantly varying with time; and (4) when the stability and control derivatives are consequently varying with time as a function of airspeed, altitude, and vehicles center of gravity.

For each of such cases, one must determine a unique gain, which could employ PID, LQR, or any suitable techniques. The gains are tabulated based on parameters such as time, altitude, or Mach number. To accomplish this goal, the nonlinear vehicle differential equations are linearized at several equilibrium flight conditions over the desired flight envelope to obtain various linear models (one for each trim point).

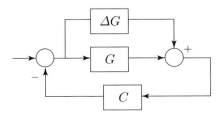

Figure 2.6: Closed-loop system with additive and multiplicative perturbations.

If the state-space representation is utilized, unique A, B, C, and D matrices are obtained. For instance, if the LQR is used for controller design, the software will be run to determine the optimal gains for each set of A, B, C, and D matrices. The optimal gains for one point in the gain schedule may be used as initial stabilizing gains in the algorithm for the next point.

2.5.4   ROBUST CONTROL

The robust control approach belongs to the family of model-based design methods that can handle the parametric uncertainty and un-modeled dynamics. The linear robust control technique (known as H_∞) can be applied to any linear system, either Jacobian or feedback linearized. In this approach, disturbances, noise, and uncertainty; ΔG (see Figure 2.6) are considered. Furthermore, an optimization technique is employed to minimize the infinity norm of the error transfer function. Consider a system described by the state-space equations:

$$\dot{x} = Ax + B_1 w + B_2 u$$
$$z = C_1 x + D_{12} u \tag{2.13}$$
$$y = C_2 x + D_{21} w.$$

The desire is to design the feedback control $u = K(s)y$, such that $\|T_{zw}(s)\|_\infty < \gamma$ for a given positive number γ. Note that γ is a function of the maximum singular value of the unstructured uncertainty (in fact., $\overline{\sigma}\left[\|\Delta G\|_\infty\right] = \frac{1}{\gamma}$). The controller (solution) is given (Ref. [39]) by the transfer function:

$$K(s) = -F(sI - \hat{A})^{-1} ZL, \tag{2.14}$$

where

$$\hat{A} = A + \frac{1}{\gamma^2} B_1 B_1^T X + B_2 F + ZLC_2 \tag{2.15}$$

and

$$F = -B_2^T X, \qquad L = -YC_2^T, \qquad Z = \left(1 - \frac{1}{\gamma^2} YX\right)^{-1}, \tag{2.16}$$

where X and Y are solutions of pairs of Algebraic Riccati equation (AREs). The closed-loop transfer function matrix $T_{zw}(s)$ from the disturbance w to the output z is given by:

$$T_{zw}(s) = G_{11} + G_{12} K (I - G_{22} K)^{-1} G_{21}, \tag{2.17}$$

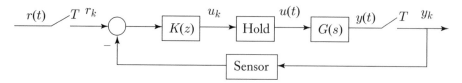

Figure 2.7: Digital control system.

where

$$G(s) = \begin{bmatrix} 0 & D_{12} \\ D_{21} & 0 \end{bmatrix} + \begin{bmatrix} C_1 \\ C_2 \end{bmatrix} (sI - A)^{-1} (B_1, B_2) = \begin{bmatrix} G_{11} & G_{12} \\ D_{21} & G_{22} \end{bmatrix}. \qquad (2.18)$$

A robust controller can handle uncertainty, disturbance, and noise. This controller demonstrated good tracking performance during a set of maneuvers.

2.5.5 DIGITAL CONTROL

In the early history of automatic flight control systems, all aspects of the flight control were analog, including the controller. With microprocessors so fast, flexible, light, and inexpensive, the control laws could be implemented in digital form. With the introduction of computers and microprocessors in the 1970s, the modern aircraft take advantage of digital control. Digital control is a branch of control theory that employs computers/microcontrollers for acting as controllers. In the digital control, a computer is responsible for the analysis and implementation of the control algorithm. A digital control system may also use a microcontroller to an application-specific integrated circuit. Since a digital device only accepts digital signal, a sampler (a kind of switch) is required to take the samples of the continuous signal. The samples are in the forms of zero (0) and one (1).

Usually, a digital control system consists of three main elements: (1) an analog-to-digital (A/D) conversion for converting analog input to digital format for the machine; (2) a digital-to-analog (D/A) conversion for converting digital output to a form that can be the input for a plant; and (3) a digital controller in the form of a computer, microcontroller, or programmable logic controller. The digital controller in implemented using a software code in a computer.

The schematic of a digital control system is shown in Figure 2.7, where z is the Z-transform variable.

The hold device is a D/A converter the discrete control samples ($K(z)$) into the continuous-time control. The sampler with sample period T is an A/D converter that takes the samples $y_k = y(kT)$ of the output of $G(s)$. In the digital control, the transfer functions are in z-domain (i.e., discrete). In a discrete (digital) system, the Laplace transform is replaced with the z-transform. The relationship between variable s and variable z is:

$$z = e^{sT}, \qquad (2.19)$$

where T is the sample rate (e.g., 0.01 sec). The approximation of the exponential function is:

$$e^{sT} \approx \frac{1 + sT/2}{1 - sT/2}. \tag{2.20}$$

This is referred to as bilinear transformation or Tustin's approximation. Inverting this transformation yields:

$$s = \frac{2}{T} \frac{z - 1}{z + 1}. \tag{2.21}$$

In the digital control, an approximate discrete equivalent of every transfer function (plant, sensor, and controller) are obtained by this transformation technique. As the sampling rate (the average number of samples, T) is higher, the approximation becomes more accurate. Reference [38] presents the analysis and design of digital control systems. Software packages such as Matlab[1] is recommended for the simulation of digital control system.

2.6 CONTROL SYSTEM DESIGN PROCESS

In this section, the overall control system design process is described. In general, the primary criteria for the design of a control system are as follows: (1) manufacturing technology; (2) required accuracy; (3) stability requirements; (4) structural stiffness; (5) load factor; (6) flying quality requirements; (7) maneuverability; (8) reliability; (9) life-cycle cost; (10) aircraft configuration; (11) stealth requirements; (12) maintainability; (13) communication system; (14) aerodynamic considerations; (15) processor; (16) complexity of trajectory; (17) compatibility with guidance system; (18) compatibility with navigation system; and (19) weight.

Among the four subsystems of an autopilot, the design of the control subsystem is the most challenging one. The guidance and navigation subsystems feed the control system to provide a successful aircraft flight mission. In general, the design process begins with a trade-off study to establish a clear line between stability and controllability requirements and ends with optimization.

During the trade-off study, two extreme limits of flying qualities are examined and the borderline between stability and controllability is drawn. For instance, a fighter aircraft can sacrifice the stability to achieve a higher controllability and maneuverability. Then, an automatic flight control system may be employed to augment the aircraft stability. In the case of a civil aircraft, the safety is the utmost goal; so the stability is clearly favored over the controllability.

The results of this trade-off study will be primarily applied to establishing the most aft and the most forward allowable location of aircraft center of gravity. Three roll control, pitch control, and yaw control are usually designed in parallel. Then the probable cross coupling between three controls is studied to ensure that each control is not negating controllability features of the aircraft in other areas. If the cross-coupling analysis reveals an unsatisfactory effect on any control plane, one or more control systems must be redesigned to resolve the issue.

---

[1]www.mathworks.com/

Flight control systems should be designed with sufficient redundancy to achieve two orders of magnitude more reliability than some desired level. In general, the control system performance requirements are: (1) fast response; (2) small overshoot; (3) zero steady-state error; (4) low damping ratio; (5) short rise time; and (6) short settling time. If the overshoot is large, a large load factor will be applied on structure (due to an increase in acceleration).

In a lightly stable aircraft, or a complete unstable aircraft, the stabilization of the aircraft is another control system requirement. This additional and necessary stabilization may make the control system design a more challenging problem.

Consider an aircraft, which its heading is controlled by a regular rudder position in the presence of a crosswind. In this system, the heading that is controlled is the direction the airplane would travel in still air. The autopilot will normally correct this heading, depending on the crosswinds, so that the actual course of the airplane coincides with the desired path. Another control included in the complete aircraft control system controls the ailerons and elevators to keep the airplane in level flight. Hence, the design of a heading control system involves the simultaneous application of all three control surface.

In FAR 23 [29], one level of redundancy for control system is required (i.e., power transmission line). The lines of power transmission (i.e., wire and pipe) should not be close to each other, should not be close to fuel tanks, and should not be close to hydraulic lines. In most Boeing aircraft, there are three separate hydraulic lines. If there is a leak in the hydraulic lines or if the engines get inoperative, there is an extra hydraulic system, which is run independently. For example, the transport aircraft Boeing 747 has four hydraulic systems. These design considerations provide a highly safe and reliable aircraft.

The main element in a control system is the controller. The major requirements to be accomplished in the design of the controller are: system stability; reference tracking; disturbance rejection; noise attenuation; control energy reduction; robust stability; and robust performance. Not all dynamic models have the ability to fully satisfy all of the requirements. Typical deficiencies of a system and the suitable compensator to compensate are tabulated in Table 2.2.

Figure 2.8 exhibits an Airbus 380 which has a maximum take-off mass of 575,000 kg and a wingspan of 78 m. This aircraft has a cruising speed of 903 km/h, a range of 14,800 km, and service ceiling of 13,100 m. It is equipped with a modern digital AFCS with a number of modes including yaw damper, Mach hold, altitude hold, vertical speed hold, approach localizer hold, heading angle hold, glide slope hold, automatic landing, automatic flare, and VOR hold modes.

Based on existing handling qualities for general aviation aircraft, and also on past experiences with aircraft flight dynamics, the following requirements are suggested in the design process: (1) stability of the overall system (minimum requirement); (2) output (or state tracking) performance; (3) accuracy command to response; and (4) minimization of a certain performance index (to follow). Specific design requirements (step response specifications) are: (1) overshoot <5%; (2) steady state error <1%; (3) rise time <1 sec; (4) settling time <3 sec; and (5) minimal cross-channel coupling.

Table 2.2: Typical deficiencies of a system and a recommended compensator

No	Deficiency of a Dynamic System	Necessary Compensator
1	Steady state error is not zero	Proportional-Integral (PI)
2	Large overshoot; long rise time; long settling time; low bandwidth	Proportional-Derivative (PD)
3	Steady state error is not zero; large overshoot; large rise time	PID Controller
4	Slow response	Lead-lag (Phase-lead)
5	Fast response	Lag-lead (Phase-lag)
6	Low damping ratio (too much oscillation)	Rate feedback
7	Sensitive to noise and disturbance	Wash-out filter
8	Unwanted mode	Zero-pole cancellation
9	A pole is not on the desired location	Pole placement
10	System response is not optimum	Quadratic Optimal Regulator
11	System's dynamic model includes uncertainty	Robust control

Figure 2.8: Airbus 380 (courtesy P. Loos at *French Wikipedia*).

The AFCS design process has an iterative nature and begins with the design requirements. Three major sections of the control system, guidance system, and navigation system are designed in parallel. In designing these systems, three laws (control law, guidance law, and navigation law) must be selected. Moreover, each of these three systems require design or a selection of a number of respective equipment.

2.7   QUESTIONS

2.1.   What are the primary criteria for the design of a control system?

2.2. Define transfer function.

2.3. What is the typical format of a state-space representation of a dynamic system?

2.4. Define the time constant.

2.5. An actuator with a first-order system model has a time constant of 0.1 sec. What is its transfer function?

2.6. Name four control laws.

2.7. Name four basic elements of a closed-loop system.

2.8. Name three nonlinearities of a control system.

2.9. What are the two conventional controller design tools/techniques?

2.10. Write the typical mathematical model for a regular actuator for control surfaces.

2.11. What is a typical value for the time constant of an aircraft actuator?

2.12. Define longitudinal control.

2.13. Define lateral control.

2.14. Define directional control.

2.15. Describe the flight envelope.

2.16. Name typical aircraft measurement devices (sensors).

2.17. Name three conventional control surfaces.

2.18. Describe PID Controller.

2.19. Describe optimal control

2.20. Describe robust control.

2.21. Describe digital control.

2.22. What do A/D and D/A stand for?

2.23. What type of compensator is recommended for a control system, where the system is sensitive to noise and disturbance?

2.24. What type of compensator is recommended for a control system, where the steady state error of the response to a unit step input is not zero?

2.25. What type of compensator is recommended for a control system, where the steady state error of the response to a unit step input is not zero, it has a large overshoot; and a large rise time?

2.26. What type of compensator is recommended for a control system, where the system's dynamic model includes uncertainty?

2.27. What type of compensator is recommended for a control system, where the system's response is not optimum?

2.28. What is the torque range of an electric motor as servo?

2.29. What is the force range of a hydraulic actuator as servo?

2.30. What are four types of electric motors used as servos?

2.31. Discuss features of a servo-motor.

2.32. Why is a saturation block needed in modeling control surfaces in Simulink?

2.33. Why can a fixed body have up to six degrees of freedom? What are they?

2.34. Draw the block diagram for the aircraft flight control system with conventional control surfaces.

2.35. Describe gain scheduling control technique.

2.36. What does MEMS stand for?

C H A P T E R   3

Attitude Control Systems

3.1   INTRODUCTION

In general, a closed-loop (i.e., negative feedback) control system tends to provide four functions: regulating, tracking, stabilizing, and improving the plant response. The regulating function here is referred to as the **hold function**, and such a system is referred to as the attitude control system. However, the tracking function is referred to as the navigation function, and such system is referred to as the flight path control systems. Here, the negative feedback is employed to regulate the output of the aircraft, that is, to hold the output constant at a "reference attitude" or a "set-point."

Automatic control systems are widely used for maintaining the attitude angles of an aircraft, or for changing an aircraft's attitude to a new commanded value. **Attitude control** (hold) system has extensive employment in long flights for conducting the earliest and most fundamental automatic flight control. They form the essential functions of an AFCS, which allow an aircraft to maintain any desired specified orientation in flight. The autopilot modes in this chapter are primarily to maintain attitude, altitude, airspeed, vertical speed, and heading. There is a group of hold functions performed by **attitude control** systems. A few regularly controlled (i.e., held) aircraft attitudes are: altitude hold; Mach hold; pitch attitude hold; bank angle hold; and turn rate hold.

The **attitude hold**—the most common AFCS function—is sometimes referred to as a control wheel steering mode. Hence, an unattended flight operation of an aircraft is possible, and no pilot input is needed.

Often, aircraft lateral and directional motions are coupled. Thus, there are two basic groups of hold functions: longitudinal hold functions and lateral-directional hold functions. In the longitudinal plane, there are primarily five functions: (1) pitch attitude hold; (2) altitude hold; (3) control wheel steering mode; (4) speed/Mach hold; and (5) vertical speed hold. However, in the Lateral-directional mode, three hold functions are dominant: (1) bank angle hold, or wing leveler; (2) heading angle hold; and (3) turn rate mode at constant speed and altitude. A summary of basic AFCS hold functions is demonstrated in Table 3.1.

Stability augmentation systems (see Chapter 5) often form the inner loops of attitude control systems; the attitude control systems then form the inner loops for the path control systems. In general, attitude control systems are often **more complex** in their operation than stability augmentation systems, since typically more control surfaces are employed.

Table 3.1: Basic AFCS hold functions

Longitudinal	Lateral-Directional
Pitch attitude hold	Bank angle hold (wing leveler)
Altitude hold	Heading angle hold
Control wheel steering mode	Turn rate hold mode at level flight
Speed/Mach hold	
Vertical speed hold	

The gain scheduling as a means of maintaining the same closed-loop performance over as much of the flight envelope as possible is recommended. To accomplish this, the nonlinear aircraft equations of motion are linearized at several trim points over the desired flight envelope to obtain state-variable models with different A and B matrices. Then the controller design is repeated for those different models.

3.2   CRUISING FLIGHT REQUIREMENTS

The UAV cruise control is assumed to be the easiest control, while having several alternatives. The longitudinal hold function is an efficient control method of cruising flight for a long period. Due to the burning fuel, the weight of aircraft at the beginning of the flight and at the end of flight is significantly different (usually about 20%). However, for a long endurance UAV such as Global Hawk, the change in the UAV weight is as much as 50%. At any weight and altitude, the lift force must be equal to aircraft weight in a straight-level flight:

$$W = L = \frac{1}{2}\rho V^2 S C_L. \tag{3.1}$$

Equation (3.1) has four independent parameters: (1) UAV weight (W); (2) airspeed (V); (3) altitude or its corresponding air density (ρ); and (4) angle of attack (α), or its associated lift coefficient (C_L). Since the fuel is consumed during flight, the aircraft weight is constantly decreased during the flight. In order to maintain a level flight, we have to decrease the lift as well. Of the many possible solutions, only three alternatives are more practical and usually examined. In each case, two flight parameters will be held constant throughout cruise. The three options of interest for continuous decrease of the lift during a cruising flight are (Figure 3.1):

1. decreasing flight speed (constant-altitude, constant-lift coefficient flight);

2. increasing altitude (constant-airspeed, constant-lift coefficient flight); and

3. decreasing angle of attack (constant-altitude, constant-airspeed flight).

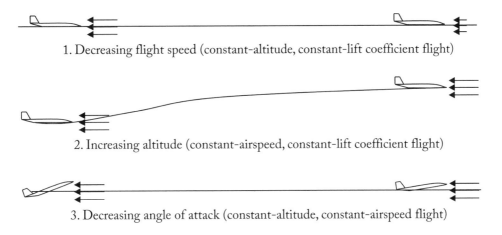

1. Decreasing flight speed (constant-altitude, constant-lift coefficient flight)

2. Increasing altitude (constant-airspeed, constant-lift coefficient flight)

3. Decreasing angle of attack (constant-altitude, constant-airspeed flight)

Figure 3.1: Three options of interest for a continuous decrease of the lift during cruise.

For each flight program, a unique controller should be designed and implemented. In the first flight program, the velocity must be reduced with the same rate as the aircraft weight is decreased. In the second solution, the air density must be decreased; in another word, the flight altitude must be increased. The third option offers the reduction of aircraft angle of attack, i.e., the reduction of lift coefficient.

In terms of AFCS application, the first flight program is applied through throttle, and the third option is implemented through stick/yoke/wheel. In the second option, no action is needed by the AFCS; the aircraft will gradually gain height (climbs).

Based on the safety regulations and practical considerations, the second flight program is the option of interest for majority of aircraft. In general, when flight is conducted under the jurisdiction of Federal Aviation Regulations, the accepted flight program is mainly the constant altitude-constant airspeed flight program.

In the first flight program, for the view-point of autopilot control, there are three draw-backs to this flight program. (1) It has the need to continuously compute the airspeed along the flight path and to reduce the throttle setting accordingly. (2) By reducing the airspeed, the flight times increases. (3) This is a fact that air traffic control rules require "constant" true airspeed for cruise flight, currently constant means ±10 knots. The good news is that, the autopilot has solved part of this problem and performs continuous calculations.

The second flight program is commonly referred to as *cruise-climb* flight. In this option, the air density will be automatically reduced as the aircraft weight is decreased. No autopilot intervention is necessary. Therefore, cruise-climb flight requires

The two modes of cruise operation discussed **here** are referred to as "Mach hold" and "altitude hold." These modes are generally employed during a cruising flight, and one or the other is selected depending on the flight mission. In terms of AFCS mode, the first and second

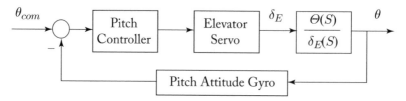

Figure 3.2: Block diagram of a pitch-attitude control system.

flight operations are controlled through the **pitch attitude hold mode**. However, the second and third flight operations are controlled through the **altitude hold mode**. No computations or efforts by the pilot.

3.3   PITCH ATTITUDE HOLD

The pitch-attitude hold mode is among one of the first AFCS modes used to help pilots in constantly controlling pitch attitude/angle in turbulent air. It is a cruise/climb control function, which is frequently used when the vehicle is in wing-level flight. Pitch attitude control systems have traditionally involved the use of elevator only as the control input. It is impossible to control the flight path angle without simultaneous control of the airspeed. Pitch attitude of the aircraft is measured by a vertical gyro. A vertical gyro is a two-degree-of-freedom gyro constructed so that the angular momentum vector (direction of the spin axis) is along the local vertical.

The controlled variable is the pitch angle (θ), which is the sum of climb angle (γ) and angle of attack (α):

$$\theta = \alpha + \gamma. \tag{3.2}$$

When the elevator is deflected or throttle setting is varied, pitch angle is changed; which implies both angle of attack and climb angle are affected. In a longitudinal control, angle of attack and climb angle are coupled; hence, with a single input, a single output will be the pitch angle. This AFCS hold function is also suitable in a climbing flight for keeping the climb angle constant.

Figure 3.2 illustrates the functional block diagram for the pitch attitude hold control system. One feedback (pitch angle) is required and the measurement device for this flight parameter is an attitude gyroscope. The pitch-angle-to-elevator-deflection transfer function ($\frac{\theta(s)}{\delta_E(s)}$) is a standard longitudinal transfer function, and can be found in flight dynamics textbooks such as [1, 2]. An elevator actuator may be modeled with a first-order transfer function. For instance, it can be modeled by a single 0.1 sec lag (i.e., $\frac{10}{s+10}$).

Using short period approximation, the pitch-angle-to-elevator-deflection transfer function is obtained [2] as:

$$\frac{\theta(s)}{\delta_E(s)} = \frac{\left(M_{\delta_E}U_o + Z_{\delta_E}M_{\dot{\alpha}}\right)s + M_\alpha Z_{\delta_E} - Z_\alpha M_{\delta_E}}{sU_o\left\{s^2 - \left(M_q + \frac{Z_\alpha}{U_o} + M_{\dot{\alpha}}\right)s + \left(\frac{Z_\alpha M_q}{U_o} - M_\alpha\right)\right\}}, \tag{3.3}$$

where U_o is initial trim speed, $Z_\alpha, M_\alpha, M_q, M_{\dot{\alpha}}$ are dimensional stability derivatives, and M_{δ_E} and Z_{δ_E} are dimensional control derivatives.

Pitch attitude is one of the effective flight variables involved in both short and phugoid (long-period) modes. In designing the pitch attitude hold system, as the controller gain is increased, the short period mode roots are driven toward the imaginary axis in root locus. This implies that, as the gain is increased, the aircraft becomes longitudinally dynamically less stable. Note that the controller does not hold the climb angle constant because the angle of attack changes with time due to fuel burn and decreasing vehicle weight. The steady state error can be removed by including an integral term in the control law. The pitch attitude feedback will improve the phugoid damping significantly.

There are some advantages to move the controller from forward loop to the feedback loop. For instance, as the feedback controller gain, is increased, the aircraft's short period frequency will also increase, although its damping ratio decreases; however, the damping ratio of the phugoid, increases. In general, feedback of pitch attitude causes the damping of the phugoid mode to increase at the expense of the damping of the short period mode. The loss of short period damping to augment the phugoid damping will result in a rather unsatisfactory dynamic response because the stability margins have been degraded.

If the engine thrust is increased, the angle of attack tends to decrease, and the vehicle will climb. As the aircraft weight decreases (due to fuel burn), the angle of attack should be decreased to reduce the lift; otherwise, it causes a gradual climb.

If the preset is the climb angle, it will level out as air density is gradually decreased. This consequently will increase the angle of attack. The pitch-attitude hold is normally employed as the inner loops for other autopilot mode such as altitude-hold and automatic landing. The pitch controller will guarantee a zero steady-state error, with a desired transient response.

An *attitude gyro* provides an error signal proportional to the deviation from a preset orientation in inertial space. An inner-loop pitch rate ($Q = d\theta/dt$) feedback can be added as the second feedback to provide a better short-period damping. The new control system with two feedbacks, which requires another sensor, is illustrated in Figure 3.3. In this control system, the pitch angle is measured with a regular attitude (i.e., vertical) gyro, while the pitch rate (Q) is measured by a rate gyro. There is an integrator ($1/s$) to derive the pitch attitude from pitch rate in the outer loop. The pitch rate is measured by a *rate gyro*.

This technique is augmenting the longitudinal stability, and will be providing a steadier flight. Two independent controllers (K_1, K_2) are changing the natural frequency and damping ratio of the pitch dynamics independently.

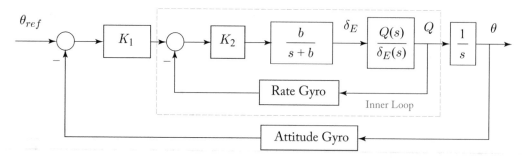

Figure 3.3: Block diagram of pitch angle control system with two feedbacks.

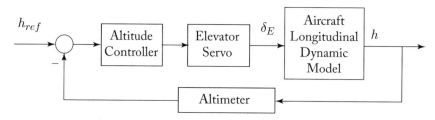

Figure 3.4: Elements of a basic altitude control system.

3.4   ALTITUDE HOLD

In an **altitude hold** mode, the aircraft's flight path angle (and thus, its altitude) is controlled by the elevator. Moreover, the airspeed or Mach number is controlled by use of the engine throttle. In this mode, one feedback (altitude) is needed, which is measured by an altimeter (as the sensor). The measurement device is an altimeter (either thorough GPS, pitot-tube, or radar altimeter). The structure of an altitude control system is very similar to rate of climb (or rate of descent) control system. In the pilot application, altitude button (ALT) of AFCS is pressed, when aircraft reaches the desired altitude.

Figure 3.4 illustrates the elements of a basic altitude hold (control) system. The purpose of the altitude hold autopilot is to hold altitude at a desired height during specified flight phases (e.g., cruise). For maintaining the constant cruise altitude, either of elevator or throttle may suffice. One option is to keep the throttle contestant (engine thrust). Thus, the elevator is employed to decrease the angle of attack as the aircraft weight is decreased. The pitch controller could be as simple as PID or to use more complicated ones. In either case, the aircraft will maintain (hold) the constant altitude through a longitudinal control system.

In designing the controller, you may begin with the approximate altitude-to-elevator-deflection transfer function. Then, the controller may be revised, by employing the complete transfer function. To derive altitude-to-elevator transfer function, we begin with relationship

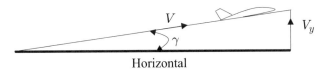

Figure 3.5: Flight path geometry for a climbing flight.

(see Figure 3.5) between the vertical speed (V_y) or rate of climb ($\dot{h}$), with aircraft speed (V) and climb angle (γ):

$$\dot{h} = V_y = V \sin(\gamma). \tag{3.4}$$

The climb angle (γ) is often small (say less than 15°). Hence, this equation can be linearized:

$$\dot{h} = V\gamma. \tag{3.5}$$

By applying Laplace transform, we will obtain:

$$sh = V\gamma(s). \tag{3.6}$$

When elevator (δ_E) is considered as the input, the following transfer function is derived:

$$\frac{h(s)}{\delta_E(s)} = \frac{V}{s} \frac{\gamma(s)}{\delta_E(s)}. \tag{3.7}$$

Using the relation between climb angle and the pitch angle (Equation (3.2)), the following is derived:

$$\frac{h(s)}{\delta_E(s)} = \frac{V_1}{s} \left[\frac{\theta(s)}{\delta_E(s)} - \frac{\alpha(s)}{\delta_E(s)} \right], \tag{3.8}$$

where V_1 is the initial trim airspeed. Two pitch-angle-to-elevator-deflection ($\frac{\theta(s)}{\delta_E(s)}$) and angle-of-attack-to-elevator-deflection ($\frac{\alpha(s)}{\delta_E(s)}$) transfer functions are standard longitudinal transfer functions, and can be found in many flight dynamics textbooks such as [2, 5] and [6]. Using short period approximation, these two transfer functions are obtained [2] as:

$$\frac{\alpha(s)}{\delta_E(s)} = \frac{Z_{\delta_E}s + M_{\delta_E}(U_1 + Z_q) - M_q Z_{\delta_E}}{U_1 \left\{ s^2 - \left(M_q + \frac{Z_\alpha}{U_1} + M_{\dot{\alpha}} \right)s + \left(\frac{Z_\alpha M_q}{U_1} - M_\alpha \right) \right\}}. \tag{3.9}$$

The altimeters tend to have a built-in lag, the magnitude of which depends on the type and its quality. Barometric altimeters and pitot-tubes have a longer lag than GPS-based altimeters. In all cases, they can be modeled by a first-order transfer function with a given time constant.

The elevator actuator may also be modeled with a first-order transfer function. Figure 3.6 demonstrates the functional block diagram of altitude hold system, where a and b are break

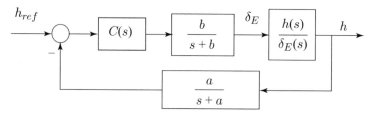

Figure 3.6: Block diagram of altitude hold system.

frequencies of altimeter and elevator actuator, respectively. In time domain, the parameters $1/a$ and $1/b$ are time constants of altimeter and elevator actuator, respectively. Various control laws may be utilized to design the controller, and to determine the controller transfer function; $C(s)$.

An accelerometer can also be used to control the altitude too. Due to the coordinate system, a_z is taken as positive downward; hence altitude (h) may be determined by double integration of the vertical acceleration (i.e., in z axis),

$$h = \frac{-1}{s^2}a_z.\tag{3.10}$$

The sign at the summing point for the feedback signal must be negative too. A specialized AFCS for military aircraft or missiles is a terrain-following terrain-avoidance (TFTA) autopilot. This flight control system often utilizes a radar carried underneath the aircraft/missile to measure the height from the Earth's surface. The radar provides guidance signal to control system in order to fly at a constant altitude (e.g., 100 ft) above the ground at constant speed. The terrain-following autopilot is presented in Chapter 4.

3.5   MACH HOLD

Cruising at a constant Mach number is a feature required by modern high subsonic transport aircraft to minimize the fuel cost while optimizing the flight time. The Mach hold mode of operation is desirable for long-range operation; however, there are many cases when it is required by regulations to fly at a constant altitude too. In an aircraft equipped with an autopilot, the cruise control—including Mach hold—will be implemented by the autopilot. After establishing the desired cruise airspeed, the pilot simply engages the Mach-hold mode (or constant-airspeed mode) on the autopilot.

Some older autopilots tend to oscillate around the target airspeed/Mach-number and constantly pitch up and down while trying to cruise a specific airspeed. This negatively impacts the passenger comfort; hence, many pilots avoid speed hold mode, especially at higher altitudes.

In the Mach hold mode, the aircraft is commanded to fly at a constant Mach number by automatically controlling the flight path angle and airspeed through the elevator and throttle. For this mode of operation, the aircraft is first trimmed to fly straight and level and the engine

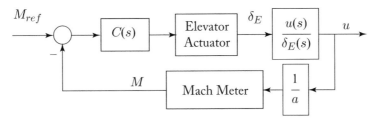

Figure 3.7: Block diagram of Mach hold mode using elevator.

power is adjusted to yield the desired Mach number. Then, the Mach hold mode of the flight control system is engaged.

As the aircraft cruises, the weight of the aircraft decreases (due to fuel consumption). If the pilot applies no input, the speed and/or altitude tend to increase (the aircraft will slowly climb). The increase in speed is sensed by the control system, and corrected for either by: an up-elevator signal or reducing the throttle. The throttle reduction is causing the aircraft to slow down; while the up-elevator signal is causing the aircraft to nose down. There is another option of interest, which uses both elevator and throttle (case 3).

In case one (using elevator only), there is only one feedback (speed/Mach-number) which is measured (Figure 3.7) by a Mach meter. There is only one controller ($C(s)$) and one control input (i.e., elevator). The net result of operation in the Mach hold mode is that the aircraft is made to climb slowly. The reason is that the angle of attack (i.e., lift coefficient) is held constant which is be compensated by lowering the air density (i.e., increasing altitude). This is a simple/basic automatic flight control function, and was mainly used in earlier subsonic transports. However, due to cruise-climb nature of this flight operation, this is not desirable by FAA, since it is not aligned with the airworthiness regulations. The throttle servo has a typical transfer function of $\frac{a}{s+a}$ where "a" is about 20 for a transport aircraft. The value 20 for "a" implies a servo time constant of 1/20 or 0.05 sec or 50 msec.

In the second case (using throttle only), the net result of operation in the Mach hold mode is that the aircraft maintains Mach number while gains height. There is only one feedback (speed/Mach-number) which is measured (Figure 3.8) by a Mach meter, and only one controller ($C(s)$) and one control input (i.e., engine throttle; δ_T). The auto-throttle adjusts the fuel supply to the engine(s). The Mach number is the airspeed divided by the speed of sound (a):

$$M = \frac{u}{a}. \tag{3.11}$$

The engine-thrust-to-throttle-setting transfer function ($\frac{T(s)}{\delta_T(s)}$) can be frequently expressed as a gain (lb/deg or N/deg). As indicated in Equation (1.16), the airspeed is mainly a function of engine throttle, aircraft weight, pitch angle, and drag. The airspeed-to-throttle-setting transfer

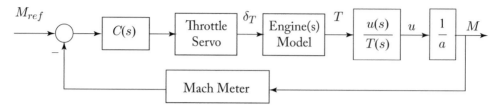

Figure 3.8: Block diagram of Mach hold mode using auto-throttle.

function ($\frac{u(s)}{T(s)}$) can be expressed as:

$$\frac{u(s)}{T(s)} = \frac{(k+s)\left(b_2s^2 + b_1s + b_o\right)}{a_4s^4 + a_3s^3 + a_2s^2 + a_1s + a_o},$$

(3.12)

where coefficients are derived by applying the following three changes into Equation 5.33 in Reference [2]: (1) replace X_{δ_e} with $1/m$; (2) replace Z_{δ_e} by zero; and (3) replace M_{δ_e} with z_T/I_{yy}. The parameter z_T is the vertical distance between thrust line and aircraft center of gravity and m represents the aircraft mass.

The thrust derivative depends on the type of engine, prop efficiency (if any), altitude, and airspeed. To have a more accurate block diagram, it is recommended to insert a first order transfer function to engine model for its time lag. The aircraft responds to this input with a change in speed as the output. The throttle servo and engine response can be modeled by a single 5 sec lag (i.e., $\frac{0.2}{s+0.2}$).

The auto-throttle is a simple/basic automatic flight control function, and was mainly used in earlier subsonic transports. However, due to cruise-climb nature of this flight operation, this is not desirable by FAA, since it is not aligned with the airworthiness regulations.

In the third case (using both elevator and throttle), the net result of operation is that the aircraft maintains both altitude and Mach number. There are two feedbacks (Figure 3.9): (1) speed/Mach-number, which is measured by a Mach meter; and (2) altitude, which is measured by an altimeter. Hence, there will be two controllers, one for elevator, and one for engine throttle.

Another variant for the Mach hold mode (option 4), is to add an inner loop (Figure 3.10) to control the pitch rate. This feature will augment aircraft longitudinal dynamic stability by adjusting four poles of short-period and phugoid modes to the desired locations. The pitch rate is measured a by a rate gyro—its gain is typically a few volts per 1 deg/sec of pitch rate. A controller of the lead-lag type (i.e., $\frac{K(s+a)}{s+b}$ can be employed to control both airspeed (i.e., Mach number) and pitch rate.

The pitch-rate-to-elevator-deflection transfer function has a fourth-order characteristic equation:

$$\frac{Q(s)}{\delta_E(s)} = \frac{s\left(b_2s^2 + b_1s + b_o\right)}{a_4s^4 + a_3s^3 + a_2s^2 + a_1s + a_o}.$$

(3.13)

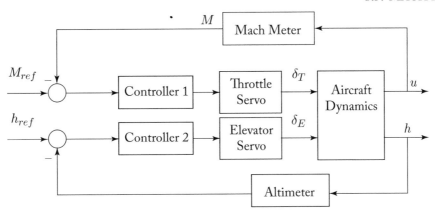

Figure 3.9: Block diagram of Mach hold mode using auto-throttle and elevator.

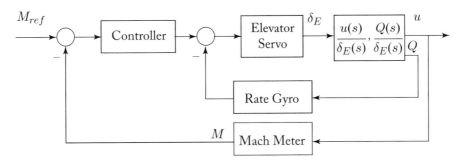

Figure 3.10: Block diagram of Mach hold mode of the AFCS with an inner loop.

Expressions for the numerator and denominator coefficients can be derived by expanding the longitudinal state-space model. However, using the short period approximation, the following transfer function with a second order characteristic equation [1] is obtained:

$$\frac{Q(s)}{\delta_E(s)} = \frac{[(U_o - Z_{\dot{\alpha}}) M_{\delta_E} + Z_{\delta_E} M_{\dot{\alpha}}] s + M_\alpha Z_{\delta_E} - Z_\alpha M_{\delta_E}}{U_o \left\{ s^2 - \left(M_q + \frac{Z_\alpha}{U_o} + M_{\dot{\alpha}} \right) s + \left(\frac{Z_\alpha M_q}{U_o} - M_\alpha \right) \right\}}. \tag{3.14}$$

The fourth variant of the Mach hold mode (Figure 3.10) is appropriate for an aircraft which experiences an undesirable Tuck mode (i.e., when the Tuck derivative (i.e., C_{m_u}) is changing sign). This derivative can be written in terms of Mach number:

$$C_{m_u} = M \frac{\partial C_m}{\partial M}. \tag{3.15}$$

If the Tuck derivative is negative (i.e., unstable), the aircraft will tend to nose down as speed increase (i.e., tuck-under effect). This behavior will become more troubling, if the elevator effectiveness is decreasing with Mach number. When the tuck derivative is negative at high subsonic

Figure 3.11: Cessna 172S Skyhawk SP.

Mach numbers, a greater low-speed static margin is required to maintain pitch stability. For this reason, the inner loop will control the pitch rate to mitigate an unstable tuck characteristic. Among various cruise control modes, one or the other selected depending on the aircraft and mission requirements. This AFCS mode can be slightly modified to develop another mode that causes the aircraft to pitch up and slow down, if the maximum speed limit (V_{NE}) is exceeded.

3.6 WING LEVELER

The earliest autopilot mode in history of aviation (in the 1930s) was the wing-leveler to reduce the pilot's need to control the bank angle in a long cruising flight. By definition, the wing-level is a zero bank angle ($\phi = 0$) or roll angle hold autopilot. Many forces will tend to disrupt a state of wings-level, such as structure's asymmetry, engine torque, atmospheric turbulence, and fuel slosh. In flying in gusty conditions, one side of the wing (left or right) will always drop. Moreover, even in the smoothest air, one side of the wing will eventually dip. So, without the autopilot intervention, the aircraft always tends to roll. This implies that no air vehicle naturally has roll stability, while they may have lateral stability.

When the autopilot holds the wings level, it provides the pilot-relief function for long flights. It will eliminate the danger of the pilot being caught unaware in a spiral motion toward the ground. Modern current single engine GA aircraft (e.g., Cessna 172; Figure 3.11) are usually equipped with an autopilot, which has at least wing leveler feature. An autopilot with only this mode is sometimes referred to as single-axis autopilot.

A means is always necessary to keep the wing level through roll control. The primary control surface (Figure 3.12) for such goal is the aileron (δ_A). Without a feedback loop, the wing level will not be activated; so a measurement device (e.g., attitude gyro) is needed to sense the

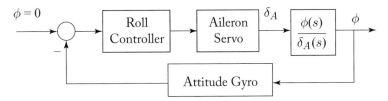

Figure 3.12: Block diagram of a basic bank angle control system.

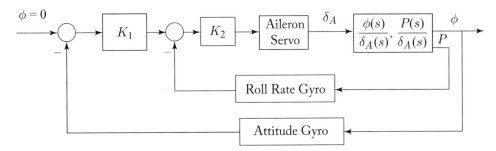

Figure 3.13: Block diagram of bank angle control system with two feedbacks.

bank angle (ϕ). To get to a zero bank angle (assuming bank angle is initially non-zero), we will need to roll and to attain a roll acceleration.

In addition to the autopilot's command of the servo (i.e., stick/yoke), there is an additional command to limit aileron saturation and over-controlling. In modern transport aircraft, a servo-position feedback loop is incorporated to limit control deflection under certain conditions.

The approximate for the bank-angle-to-aileron-deflection transfer function is modeled by a second-order system as:

$$\frac{\phi(s)}{\delta_A(s)} = \frac{L_\delta A}{s^2 - sL_p},\tag{3.16}$$

where L_p and $L_{\delta A}$ are dimensional roll damping (stability) and roll control derivatives, respectively. The zero root (free s) means that if the aircraft is disturbed from its desired bank angle, there is no inherent restoring moment unless the pilot applies aileron correction.

Another more effective method to control the bank angle is to have two feedbacks (Figure 3.13) which requires two sensors. In this control system, the bank angle is measured with a regular attitude gyro, while the roll rate (P) is measured by a rate gyro. This technique is augmenting the directional stability; and provides a steadier flight. Two independent controllers (K_1, K_2) are changing the natural frequency and damping ratio of the rolling motion independently.

The differential governing equation of motion for a pure rolling motion [9] is

$$\frac{-1}{L_P}\frac{dP}{dt} + P = \frac{L_{\delta A}}{L_p}\delta_A. \tag{3.17}$$

Applying Laplace transform, and deriving the transfer function will yield:

$$\frac{-1}{L_P}sP + P = \frac{L_{\delta A}}{L_p}\delta_A. \tag{3.18}$$

$$\frac{P(s)}{\delta_A(s)} = \frac{\dfrac{L_{\delta A}}{L_p}}{\dfrac{-1}{L_P}s + 1} = \frac{-L_{\delta A}}{s - L_P}. \tag{3.19}$$

Recall that for a fixed-wing UAV, the damping derivative L_P is negative, so the roll-rate-to-rudder is a stable system. This transfer function, plus the bank angle transfer function will be employed in the development of the bank angle control system.

Since most aircraft has strong yaw-roll coupling, the roll-angle-to-aileron feedback should be considered in any directional control system design. When both roll rate feedback and bank angle feedback are simultaneously used, there is good control over the position of the closed-loop poles in designing controllers. When the bank angle feedback loop is used, the spiral pole moves to the left, and the Dutch roll poles move to the right of the s-plane. This implies that Dutch roll mode will become unstable, while the spiral mode gets stable.

3.7   TURN COORDINATOR

3.7.1   GOVERNING EQUATIONS OF A COORDINATED TURN

Turn is a flight maneuver, which leads to a change in the heading angle (Ψ); it is usually performed by simultaneous rolling and yawing motions (so, bank-to-turn). In rare cases (e.g., some missiles), a turn only involves a yawing motion (skid-to-turn). The most desired type of a level turn is the one which is coordinated; which is defined as a turn which the lateral acceleration is zero. It has a few advantages including a constant radius, no lateral load factor, and better environment for mechanical instruments.

A coordinated turn is [2] one in which the aerodynamic side-force (F_{Ay}) and the lateral acceleration (a_y) are both zero. According to [3], in a coordinated turn, the sideslip angle should also be zero. The author believes that the first condition suffices for a coordinated turn (i.e., $F_{Ay} = a_y = 0$). Thus, a coordinated turn features no slip and no skid.

The result of a coordinated turn is that the turn radius (R) will be kept constant. Hence, the sideslip angle is the motion variable whose control is central to the achievement of a coordinated turn. Every modern aircraft has a turn coordinator, which has an instrument consisting of a rate gyro to indicate the rate of yaw, and an accelerometer to determine the centripetal acceleration.

In such turns, there is minimum coupling of rolling and yawing motions. Hence, the aircraft is traveling in a circular path with a constant radius and a constant airspeed (V).

Hence, the horizontal component of the lift (L) is equal to the centrifugal force, while its vertical component equals the vehicle's weight:

$$L \sin \phi = m \frac{V^2}{R} \tag{3.20}$$

$$L \cos \phi = W, \tag{3.21}$$

where ϕ is the bank angle. The turn coordinator is a lateral autopilot, which provides a body lateral load factor, n_y, to a commanded load factor, n_{yc} (often zero). There are various configurations for a turn coordinator; most includes an inner loop (yaw rate damper) and an outer loop. The outer loop uses an accelerometer feedback and has the capability of moment arm feedback, if the accelerometer is located at a different position than the UAV center of gravity.

The ratio between the lift and the aircraft weight (W) is called the load factor (n):

$$n = \frac{L}{W}. \tag{3.22}$$

The relationship [38] between turn radius, bank angle, and airspeed in a coordinated turn is:

$$R = \frac{V^2}{g \tan \phi}. \tag{3.23}$$

When the sensors in the turn coordinator remind the flight computer that the airplane is turning, the computer sends a signal to the roll servo (an electric motor, or a hydraulic jack). Then, through a bridle cable (if a mechanical system), grips one of the aileron cables. As the roll servo gently applies aileron against the turn, the flight computer monitors the progress, eventually removing the command when the turn coordinator signals that the wings are once again level. This loop works continuously, many times a second.

In a coordinated turn, the aircraft maintains a constant pitch and bank angles, while the heading angle continuously changes at a constant rate. Hence, the Eulerian roll rate ($\dot{\phi}$) and pitch rate ($\dot{\theta}$) are zero, while the Eulerian yaw rate ($\dot{\psi}$) is turn rate. The body-axes angular velocities are functions of pitch angle and bank angle:

$$P = -\dot{\psi} \sin(\theta) \tag{3.24}$$

$$Q = \dot{\psi} \sin(\phi) \cos(\theta) \tag{3.25}$$

$$R = \dot{\psi} \cos(\phi) \cos(\theta). \tag{3.26}$$

These three flight parameters can be measured and used as feedback to control a coordinated turn. However, since the pitch angle (θ) is small, the linearization yields: $\sin(\theta) = 0$, and $\cos(\theta) = 1$. Thus, for a specified turn rate, the roll rate can be neglected, and the required yaw rate can be calculated. To hold a constant altitude, the elevator should also be employed.

Table 3.2: Autopilot inner loops

No	Loop	Command Variable	Control Surface	Constraints
1	Normal acceleration command	n_z	Elevator	$-30° \leq \delta_E \leq +30°$ $-2 \leq n_z \leq +5$
2	Bank angle command	ϕ	Aileron	$-30° \leq \delta_A \leq +30°$ $-2 \leq n_z \leq +5$
3	Turn coordinator	Lateral acceleration (n_y)	Rudder	$-30° \leq \delta_R \leq +30°$ $n_y = 0$

Table 3.2 presents the inner loops of a turn coordination mode of an autopilot. In a coordinated turn, all three control surfaces (in a conventional fixed-wing UAV) are simultaneously employed. The elevator is deflected to maintain the desired angle of attack to compensate the lift. The aileron is deflected to develop the bank angle. The rudder has the primary role in transferring a regular turn to a coordinated one. In a bank-to-turn ($\phi > 0$) maneuver, due to an increase in the angle of attack, the normal load factor (n_z) will be more than one. The relation between normal load factor and the vehicle bank angle is:

$$n_z = \frac{1}{\cos(\phi)}. \tag{3.27}$$

Hence, as the bank angle is increased, the load factor is increased too.

3.7.2   BLOCK DIAGRAM

Inner-loop augmentation and control modes should be designed to improve aircraft damping in all axes. The turn coordinator frequently satisfies a commanded load factor and a commanded bank angle (the block diagram is similar to that of a wing leveler). The outer loops should be designed to provide altitude hold, and vertical flight path angle hold. Moreover, if the guidance system is involved, the guidance loops should be designed to provide proportional navigation in elevation and azimuth. For a non-fighter aircraft, the lateral acceleration (a_y or a_{lat}) is desired to be zero, which is sensed by a lateral accelerometer. To develop an easier design environment for the turn coordinator, it is recommended to install the lateral accelerometer at the vehicle center of gravity. The lateral accelerometer is very sensitive to structural vibration, so an appropriate filter must be used.

The block diagram representing a basic turn coordinator mode of an autopilot is shown in Figure 3.14. There are three feedbacks (lateral acceleration, yaw rate, and rate of yaw rate), and two sensors (lateral accelerometer and yaw rate gyro). The parameter "C" in the second loop represents the moment arm for the accelerometer. If the accelerometer is located at the vehicle's

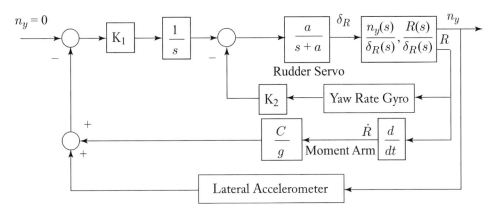

Figure 3.14: Block diagram of a basic turn coordinator.

cg, the "C" will be zero. Various controllers may be employed to satisfy the control requirements; the simplest ones are two gains. When all transfer functions are derived, the design problem statement is to determine two gains K_1 and K_2 to satisfy the turn coordination requirements.

The components of load factor in y and z directions may be expressed [5] as the function of flight variables. The normal load factor as a function of rate of angle of attack ($\dot{\alpha}$) is:

$$n_z = \frac{U_o}{g}[Q - \dot{\alpha}], \tag{3.28}$$

where U_o is the total initial airspeed. The lateral load factor (n_y) is defined as:

$$n_y = \frac{F_y}{mg}. \tag{3.29}$$

The lateral load factor as a function of yaw rate (R) and rate of angle of attack ($\dot{\beta}$) is obtained as:

$$n_y = \frac{U_o}{g}\left[\dot{\beta} + R\right]. \tag{3.30}$$

There is a relationship between speed in y-direction (v) and sideslip angle (β) as:

$$\tan(\beta) = \frac{v}{U_o}. \tag{3.31}$$

By assuming small angles, and applying the linearization technique (small perturbation theory; $\sin\beta = \beta$ and $\cos\beta = 1$), this equation may be simplified to:

$$\beta = \frac{v}{U_o}. \tag{3.32}$$

Differentiating both sides yields

$$\dot{v} = \dot{\beta} U_o. \tag{3.33}$$

Using this differential equation, the $\dot{v}$ is replaced with $\dot{\beta} U_o$:

$$n_y = \frac{1}{g} [\dot{v} + U_o R]. \tag{3.34}$$

The yaw-rate-to-rudder-deflection transfer function [5] is given by:

$$\frac{R(s)}{\delta_R(s)} = \frac{U_o N_{\delta_R} s + (N_\beta Y_{\delta_R} - Y_\beta N_\beta)}{U_o s^2 + s(U_o N_r - Y_\beta) + (Y_\beta N_r + U_o N_\beta - Y_r N_\beta)}. \tag{3.35}$$

Inserting this transfer function into Equation (3.34), the transfer function for lateral-acceleration-to-rudder-deflection is obtained:

$$\frac{n_y(s)}{\delta_R(s)} = \frac{U_o}{g} \frac{Y_{\delta_R} s^2 + s(Y_r N_{\delta_R} - Y_{\delta_R} N_r) + (Y_{\delta_R} N_\beta - N_{\delta_R} Y_\beta)}{U_o N_{\delta_R} s + (N_\beta Y_{\delta_R} - N_{\delta_R} Y_\beta)}. \tag{3.36}$$

Note that an n_y of zero means a coordinated turn.

3.7.3   AILERON-TO-RUDDER INTERCONNECT

It has long been a practice in coordinated turn to incorporate a control interconnect (sometimes referred to as cross-feed) to nullify the side force. Two effective cross-feed techniques are: **aileron-to-rudder interconnect (ARI)** and bank angle to rudder crossfeed. Both techniques are widely employed in large transport aircraft.

Figure 3.15 demonstrates the block diagram of a turn coordination system which utilizes ARI. It has three feedbacks: bank angle, roll rate, and yaw rate. Moreover, the system employs two controllers and one crossfeed gain. The presence of the wash-out filter in the crossfeed path is required to allow the aircraft to produce steady non-zero sideslip angle. Both aileron and rudder are used to coordinate the turn using an ARI technique.

The bank angle command is a function of turn rate and turn radius. In a coordinated turn, the yaw rate is [3] given by:

$$R = \frac{g \sin \phi}{V}. \tag{3.37}$$

If this equation does not hold, the aircraft will slip or skip, either of which is undesirable. For such cases, the error signal will be:

$$e_{ss} = K_c \left(R - \frac{g \sin \phi}{V} \right), \tag{3.38}$$

where K_c is the conversion factor. For a small bank angle—using small disturbance theory—Equation (3.37) may be linearized (i.e., $\sin \phi = \phi$). Thus, the command signal for yaw rate can be taken as:

$$R = \frac{g}{V} \phi = K_{cf} \phi. \tag{3.39}$$

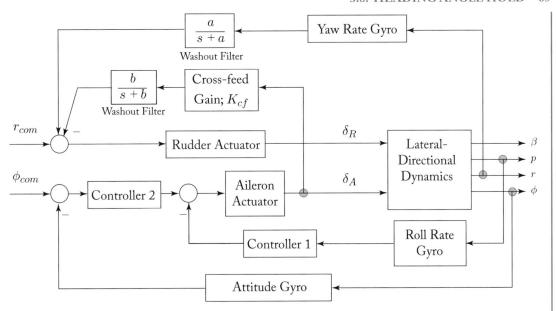

Figure 3.15: Block diagram of aileron-to-rudder interconnect system.

Such crossfeed to the rudder channel will cause a phase compensation, which will increase the Dutch roll damping. In a case of AFCS utilizing crossfeed, if failure occurs in any feedback path, the flying qualities of the aircraft are drastically impaired. For such situations, it is recommended to immediately disconnect AFCS (i.e., disconnect the bank angle signal from the rudder). In designing ARI, two controllers are designed, and a crossfeed gain is determined, such that the step response are satisfactory, so that the turn is effective.

3.8 HEADING ANGLE HOLD

The heading angle (Ψ) is defined as the angle between the aircraft direction in x-y plane and a reference line (e.g., North). The sideslip angle is the angle between the aircraft x-axis and the flight path in the x-y plane. However, in pilot terminology, the term heading is used to describe the direction an aircraft is pointing, while the course angle refers to the direction an aircraft is actually flying. The difference between course and heading angles is called the crab angle. Bearing is the angle (clockwise) between North and the flight path, or to a navigation aid station (e.g., VOR). A heading angle hold function is an AFCS mode to hold an aircraft on a desired heading. This is an important automatic navigation function of an autopilot for both military and civil aircraft.

Figure 3.16 illustrates the functional block diagram for the heading angle hold system. One feedback (heading angle) is required for this system, and the measurement device for this

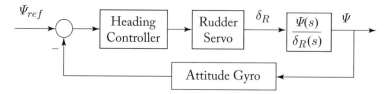

Figure 3.16: Block diagram of basic heading angle hold system.

flight parameter is often an attitude (directional) gyro (i.e., heading indicator). The heading indicator is a mechanical instrument designed to facilitate the use of the magnetic compass through a gyroscope. Some aircraft employ a horizontal situation indicator (HSI) which receives a magnetic north reference from a magnetic transmitter (i.e., **magnetometer**). Moreover, a magnetic compass or GPS device will create a reference command (Ψ_{Ref}). The AFCS should be able to track the selected radio course, with automatic crosswind correction.

The heading-angle-to-rudder-deflection transfer function ($\frac{\psi(s)}{\delta_R(s)}$) is a standard directional transfer function, and can be found in flight dynamics textbooks such as [1, 2]. The rudder actuator may be modeled with a first order transfer function.

The heading-angle-to-rudder-deflection transfer function has a fifth-order characteristic equation:

$$\frac{\psi(s)}{\delta_R(s)} = \frac{b_2 s^2 + b_1 s + b_o}{s\left(a_4 s^4 + a_3 s^3 + a_2 s^2 + a_1 s + a_o\right)}. \tag{3.40}$$

Expressions for the numerator and denominator coefficients can be derived by expanding the lateral-directional state-space model. However, using approximate yawing moment equation, the following transfer function with a second-order characteristic equation [1] is obtained:

$$\frac{\Psi(s)}{\delta_R(s)} = \frac{N_{\delta R}}{s^2 - s N_r + N_\beta}, \tag{3.41}$$

where N_r and N_β are dimensional yaw stability derivative, and $N_{\delta R}$ is the dimensional yaw control derivative.

Since most aircraft has strong yaw-roll coupling, the roll-angle-to-aileron-deflection feedback should also be considered in any heading hold system design. To change the heading angle, a certain bank angle is established and held until the desired heading angle is gained. Thus, a bank angle control loop is utilized as the inner loop in a heading angle hold mode.

In a coordinated turn, the linear velocity (i.e., airspeed, V) and angular velocity (i.e., turn rate, $\dot{\Psi}$) are related via turn radius:

$$V = R\dot{\Psi}. \tag{3.42}$$

The turn radius is a function of bank angle as:

$$R = \frac{V^2}{g \tan(\phi)}. \tag{3.43}$$

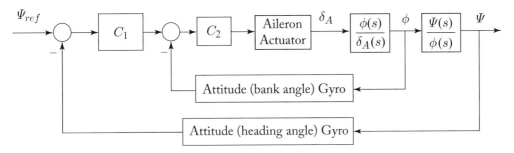

Figure 3.17: Block diagram of a heading angle hold mode with a bank angle inner loop.

By eliminating the turn radius from Equations (3.42) and (3.43), the corresponding turn rate is obtained:

$$\dot{\Psi} = \frac{g \tan(\phi)}{V}. \tag{3.44}$$

When this equation is linearized ($\tan(\phi) = \phi$) and the Laplace transform is implemented, the following transfer function is obtained:

$$\frac{\psi(s)}{\phi(s)} = \frac{g}{sV}. \tag{3.45}$$

This transfer function should be used in the forward path of the outer loop. The bank angle control loop of Figure 3.12 should be employed as the inner loop for the heading hold system of Figure 3.16. The outcome will be a control system with an inner loop and an outer loop as indicated in Figure 3.17. A vertical gyro is used for the purpose of measuring bank angle, and a directional gyro is used for the heading reference.

The C_1 is the controller for the outer loop (heading angle), and the C_2 is the controller for the inner loop (bank angle). Two controllers are designed simultaneously to obtain the desired yaw performance and gain acceptable flying qualities. The bank-angle-to-aileron-deflection transfer function ($\frac{\phi(s)}{\delta_A(s)}$) is introduced earlier in Equation (3.16).

This is one technique to control heading angle, there are other methods to hold this angle (hold the aircraft on a desired heading). It is left to the interested reader to explore other methods. For instance, turn to a new heading could be accomplished by employing a turn control, which will disengage the heading reference and command a yaw rate. When the turn is completed, the commanded yaw rate is changed to zero, and the heading reference will automatically re-engage at the new heading angle.

The use of the heading error to generate the yaw rate command can result in a high commanded yaw rate that is beyond the aircraft maneuverability (i.e., structural) limit. To avoid this outcome, a limiting circuit (i.e., saturation) must be placed in the control system. The maximum value of the commanded yaw rate is limited by parameters such as the maximum airspeed, maximum acceleration, and maximum allowable load factor.

Figure 3.18: Runway 22 and aircraft heading.

In a cruising flight, either compass or GPS may be employed as the **reference command generator** (resetting the heading reference). Recall that a compass always has an inclination with geographic north, since it aligns itself with the Earth's magnetic field. The difference between true north and magnetic north is called magnetic declination (or magnetic deviation).

However, in a landing approach, using a compass heading (not a GPS heading) is highly recommended. Airport runways are designated with reference to the magnetic north, not the geographic north. Runways are named by a number between 01 (implies 10°) and 36 (implies 360°), which is the magnetic heading of a runway in deca-degrees. Therefore, an aircraft trying to land on a runway marked as 22 (Figure 3.18) should align its fuselage nose to follow the compass when it indicates 220°. It such landing condition, the GPS may indicate a heading of 204° or 235°.

3.9   VERTICAL SPEED HOLD

Any regular flight, which begins with a take-off and ends with a landing, should include at least one climb and at least one descent. The Vertical Speed Hold mode of an AFCS is designed to automatically control the climb and descent by holding a desired vertical speed (V/S or VS). This mode is very helpful when ATC requests the pilot to maintain a specified airspeed in a

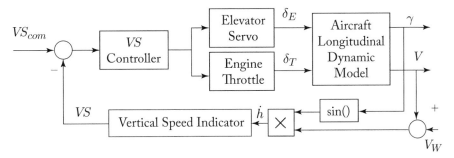

Figure 3.19: Block diagram of a basic vertical speed hold system.

climb or descent. After a take-off, the vertical speed mode can hold the vertical speed until a target altitude is reached. The autopilot is frequently engaged in the vertical speed mode for the climb to near the assigned cruise altitude. Another application is in the Flight Level Change (FCL) mode which controls pitch angle and airspeed to maintain a constant indicated airspeed. This AFCS mode can be commanded to start a fixed rate climb/descent to avoid weather during the cruise phase. In general, a decent after cruise will typically use a shallower descent than an approach glide.

The block diagram of a basic vertical speed hold system is depicted in Figure 3.19. The control system has mainly one commanded input (VS_{com}), two control variables (elevator, δ_E and engine throttle, δ_T), two outputs (ground speed, V and climb/descent angle, γ), and one feedback.

Both elevator and engine throttle simultaneously are influencing the airspeed and the climb/descent angle. If there is no wind (head wind or tail wind), the airspeed will be the same as the ground speed. The wind will affect the speed of the airplane relative to the earth—i.e., its ground speed (V_G)—however, it will not affect its speed relative to the air (V or V_A).

$$\overrightarrow{V_G} = \overrightarrow{V_A} \pm \overrightarrow{V_W}, \tag{3.46}$$

where V_A is the airspeed, and V_W denotes the wind speed. The wind speed is indicated (i.e., modeled) as an added noise to the block diagram. The Vertical Speed Indicator (VSI) measurement includes this speed when measuring the change in the static pressure.

The vertical speed is a measure of rate of climb ($+\dot{h}$), or rate of descent ($-\dot{h}$). It is (Figure 3.5) a function of aircraft ground speed (V) and climb/descent angle (γ):

$$VS = V \sin(\gamma). \tag{3.47}$$

The climb-angle-to-elevator-deflection transfer function ($\frac{\gamma(s)}{\delta_E(s)}$), airspeed-to-elevator-deflection transfer function ($\frac{V(s)}{\delta_E(s)}$), and airspeed-to-throttle-setting transfer function ($\frac{V(s)}{\delta_T(s)}$) are introduced in Chapter 4. The measurement device, VSI, can be modeled by a first-order transfer function:

$$H(s) = \frac{K_{VSI}}{\tau s + 1},$$
(3.48)

where the time constant, τ is about 1 sec. The constant K_{VSI} is a function of *VSI* configuration, which converts the static pressure change to an output signal (either mechanical or electric).

The measurement device is a VSI, which provides an indication of rate of change of altitude. This instrument is referred with other names such device are: (1) variometer; (2) rate of climb and descent indicator; (3) rate of climb indicator; (4) vertical speed indicator; and (5) vertical velocity indicator. The VSI measures the rate of change of altitude by detecting the change in air pressure (static pressure) as altitude changes. The VSI is typically connected to the static pressure source, either fuselage side surfaces or from the static port on the pitot tube. Modern high precision electronic barometric sensors provide static pressure information in a digital form.

3.10   QUESTIONS

3.1.   Name four functions of a closed-loop control system.

3.2.   Name at least three longitudinal hold functions performed by attitude control systems.

3.3.   Name at least three lateral-directional hold functions performed by attitude control systems.

3.4.   List the three options of interest for continuous decrease of the lift during a cruising flight.

3.5.   What is the primary tool for decreasing flight speed in a constant-altitude, constant-lift coefficient cruising flight?

3.6.   What is the primary tool for increasing altitude in a constant-airspeed, constant-lift coefficient cruising flight?

3.7.   What is the primary tool for decreasing angle of attack in a constant-altitude, constant-airspeed cruising flight?

3.8.   Briefly describe the features of a *cruise-climb* flight. What variable is held constant in this flight?

3.9.   Briefly describe the features of a vertical gyro.

3.10.   Draw the functional block diagram for the pitch attitude hold control system.

3.11.   Describe a model for an elevator actuator.

3.12.   What is the primary objective for the pitch attitude hold control system?

3.13.   Draw the block diagram of pitch angle control system with two feedbacks.

3.14. What is the typical measurement device for altitude control system?

3.15. Draw the block diagram of an altitude control system.

3.16. Name main elements for an altitude control system.

3.17. Define climb angle.

3.18. Explain how an altitude may be determine by via a vertical acceleration.

3.19. Draw the block diagram of Mach hold mode using elevator.

3.20. What is the typical form of an engine-thrust-to-throttle-setting transfer function?

3.21. Provide a technique to derive the airspeed-to-throttle-setting transfer function.

3.22. Draw the block diagram of Mach hold mode using auto-throttle.

3.23. Draw the block diagram of Mach hold mode using auto-throttle and elevator.

3.24. Draw the block diagram of Mach hold mode of the AFCS with an inner loop.

3.25. What is the Tuck derivative?

3.26. What is the earliest autopilot mode in history of aviation?

3.27. What is a single-axis autopilot?

3.28. Is there any autopilot function in Cessna 172?

3.29. Draw the block diagram of a basic bank angle control system.

3.30. Introduce an approximate for the bank-angle-to-aileron-deflection transfer function.

3.31. Draw the block diagram of bank angle control system with two feedbacks.

3.32. Compare the differences between features of an AFCS ahold mode when both roll rate feedback and bank angle feedback are simultaneously used, with the one with only bank angle feedback loop is used.

3.33. Define a coordinated turn.

3.34. What is the difference between a skid-to-turn and a bank-to-turn?

3.35. What are the advantages of a coordinated turn?

3.36. Define the load factor.

3.37. What variables are maintained constant during a coordinated turn?

3.38. What variable is desired to be zero during a coordinated turn?

3.39. Draw the block diagram of a basic turn coordinator.

3.40. What control derivatives are influencing the yaw-rate-to-rudder-deflection transfer function?

3.41. What stability derivatives are influencing the lateral-acceleration-to-rudder-deflection transfer function?

3.42. Why a control interconnect is incorporated in a coordinated turn?

3.43. What is the objective for an aileron-to-rudder interconnect (ARI)?

3.44. Draw the block diagram of aileron-to-rudder interconnect system.

3.45. Define the heading angle.

3.46. Define the crab angle.

3.47. Draw the block diagram of a basic heading angle hold system.

3.48. What are: (1) control input and (2) measurement device, for a basic heading angle hold system?

3.49. What stability derivatives are influencing the heading-angle-to-rudder-deflection transfer function?

3.50. Draw the block diagram of a heading angle hold mode with a bank angle inner loop.

3.51. What is the magnetic declination?

3.52. What is the heading of a runway coded 22?

3.53. Draw the block diagram of a basic vertical speed hold system.

3.54. Describe how an aircraft vertical speed is influenced by a wind speed.

3.55. Describe the mechanism for a measurement device to measure the vertical speed.

3.56. What are other names for a Vertical Speed Indicator?

3.57. In a cruising flight, what element is often used to change the airspeed?

3.58. In a cruising flight, what element is often used to change the angle of attack/pitch angle?

3.59. In a cruising flight, what element is often used to change the altitude?

3.60. Name a typical device to measure the pitch rate.

3.61. Describe briefly the technique to derive the airspeed-to-throttle-setting transfer function.

3.62. What can be implied from the free "s" in denominator of the bank-angle-to-aileron-deflection transfer function?

CHAPTER 4

Flight Path Control Systems

4.1 INTRODUCTION

In general, a closed-loop (i.e., negative feedback) control system tends to provide four functions: (1) regulating; (2) tracking; (3) stabilizing; and (4) improving the plant response. The regulating function here is referred to as the hold function, and such a system is referred to as the attitude control system (see Chapter 3). However, the tracking function is referred to as the **navigation function**, and such system is referred to as the flight path control systems. Here, the negative feedback is employed to make the output "track" (i.e., follow) a changing command of the aircraft, a "commanded attitude."

The control theory primarily deals with following a nonzero reference command signal (i.e., tracking a command), not with regulating the state near zero (i.e., hold functions), which we explored in the preceding chapter. The flight path can be tracked and controlled by feedback comparison with a precomputed reference trajectory, or with real-time trajectory generation. Hence, three systems (Figure 4.2) are working in conjunction with flight control system: (1) trajectory generation system, (2) navigation system, and (3) guidance system.

The trajectory generation system will constantly create a new reference command, while the coordinates of the vehicle are determined by navigation system at any instance. The guidance system compares the current coordinates with the desired coordinates and generates the guidance command for the control system. The control system will constantly compare the current aircraft coordinate with the commanded values and will apply corrections through control surfaces and engine throttle. In some textbooks (e.g., [2]), and in pilots' terminology, the flight path control systems are referred to as the navigation functions of AFCS.

The navigation functions are used to follow the programmed flight route. The autopilot directional navigation function steers the aircraft to keep the course deviation indicator (CDI) needle centered. A CDI is to determine an aircraft's lateral position in relation to a course to or from a radio navigation beacon.

In theory, navigation is the skill that involves the determination of position, orientation, and velocity of a moving object. More specifically, navigation is a field of study that focuses on the process of monitoring the movement of a vehicle from one place to another. The aircraft navigation system is one that determines the position of the air vehicle with respect to some reference frame (i.e., situational awareness), for example, the Earth sea level. However, the AFCS navigation functions are kinds of control modes in which the flight path is controlled. Thus, another term for the navigation function is the flight path control system. In these modes, the

Figure 4.1: Navigation function to follow a commanded trajectory.

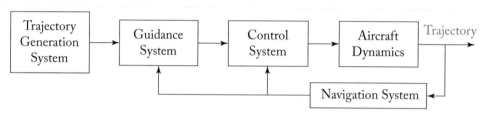

Figure 4.2: Three systems working in AFCS navigation mode.

aircraft is required to fly (Figure 4.1) from waypoint 1 (with coordinates x_1, y_1, z_1) to waypoint 2 (with coordinates x_2, y_2, z_2).

For the navigation functions, there are two groups of modes: longitudinal navigation functions and lateral-directional navigation functions. In longitudinal plane, there are primarily seven modes: cruise control; automatic flight level change; terrain following; automatic climb and descent; automatic flare control; glide slope track; and automatic landing. However, in lateral-directional mode, three hold functions are dominant: approach localizer track; VOR hold; heading tracking; tracking a series of waypoints; and turn coordination.

VOR is a type of radio navigation system for aircraft. The VOR station broadcasts a VHF radio signal, and data that allows the airborne receiver to derive the magnetic bearing from the station to the aircraft. This line of position is referred to as the "radial." Pilots utilize this information to determine their position and navigate the aircraft to their destination. A summary of two groups of navigation functions is demonstrated in Table 4.1.

The primary flight display often integrates all controls that allow modes to be entered for the autopilot. Figure 4.3 exhibits the primary flight display of a Boeing 737 MAX 8. It is observed that in the mode control panel (MCP), functions such as altitude hold (ALT HOLD), and lateral navigation (LNAV) are engaged.

The block diagram of every type of a flight path control system includes at least one controller. The final design of the controller should be verified on the complete aircraft dynamics via a simulation. In this chapter, seven longitudinal and five lateral-directional AFCS navigation functions are discussed.

Table 4.1: **Navigation functions**

No	A. Longitudinal	B. Lateral-Directional
1	Cruise control	Approach localizer tracking
2	Automatic flight level change	VOR tracking
3	Terrain following	Heading tracking
4	Automatic climb and descent	Tracking a series of waypoints
5	Automatic flare control	Turn coordination
6	Glide slope tracking	
7	Automatic landing	

Figure 4.3: Primary flight display of a Boeing 737 MAX 8.

4.2   LANDING OPERATION PROCEDURES

Due to the complexity of a landing operation, a brief description of its various phases and airport instruments are beneficial. Landing is the last phase of a normal flight, which brings back the aircraft from airborne to the ground. After a descent from a cruising altitude, landing will be the last flight operation. A landing operation begins with an approach, then flare (round out), and rotation, and will be finished with braking. The landing operation is divided [13] into various segments: approach (heading correction phase); approach (glide slope correction phase); flare (round out); touchdown; de-rotation; roll out; and deceleration. A correctly controlled approach ensures that the airspeed is appropriate, the rate of descent is slow enough, and that the flight path would terminate within the touchdown zone. After the landing, the aircraft will be taxied to the desired gate for cargo/passenger release.

The flare (rotation) is the last part of an approach and brings the fuselage nose up such that the aircraft has a high angle of attack. At this moment, aircraft speed is about 1.3 times the stall speed (V_s). During this section, simultaneous with speed reduction, the height between aircraft and runway is gradually reduced, until the main gear touches down. This section of landing is all airborne.

Airports are classified from various aspects, including: runway length; type of approach instruments; and presence of an Air Traffic Control Tower. In terms of landing approach instruments, there are two airport types: airports that are equipped with Instrument Landing System (ILS) and airports without ILS.

The airports with ILS are equipped with a number of instruments (e.g., localizer; glide slope transmitter; marker beacon; compass locators (i.e., direction indicators); and distance Measuring Equipment) and signs (e.g., approach lights; touchdown; centerline lights; and runway lights).

An ILS provides three groups of information to incoming aircraft: (1) guidance information (via localizer and glide slope); (2) range information (via outer marker and the middle marker beacons; (3) visual information(via approach lights, touchdown, and centerline lights, runway lights). In airports with ILS, the incoming aircraft should detect airport localizer beam, and adjust its heading with the localizer signal. Section 4.3 presents the control system for approach localizer tracking.

In airports without ILS, an approaching aircraft should adjust its heading with the runway direction using a compass heading and the pilot should make sure that, the compass indicator centered. Airport runways are designated with reference to the magnetic north, not the geographic north. The airport runways' headings (using two digits) are published in "Aeronautical charts." For example, an aircraft trying to land on a runway marked as 22, should align its fuselage nose to follow the compass when it indicates 220°.

The airport instrument landing system (ILS) broadcasts signals [13] to arriving instrument aircraft to guide them to the runway. The automatic approach often works by using the help from two airport instruments: a localizer, and a glide slope transmitter. These two instruments form the horizontal (directional) and longitudinal guidance system for automatic landing operation. A localizer transmits VHF signal for lateral guidance, while the glide slope transmits UHF signal for vertical guidance. There are a number of methods for automatic landing approach.

A localizer (i.e., horizontal component of ILS) is a navigation instrument, which is used to control the vehicle heading along the axis of the runway. It is a transmitter with a horizontal antenna array located at the opposite end of runway; the aircraft receives the broadcasted signal through its receiver antenna. Figure 4.4 illustrates two airport transmitters, which transmit two guidance beams to approaching aircraft, along with the geometry of landing flight path.

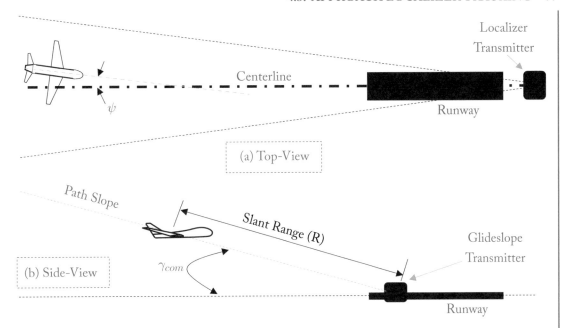

Figure 4.4: Airport transmitters transmit two guidance beams to approaching aircraft.

A glide-slope transmitter (i.e., vertical component of ILS) transmits a directional pattern to define a glide-path beam and to assist an aircraft to make an approach at a desired glide angle (γ_{com}). Its radio beam extends upward approximately 2 1/4–3 1/2° from the approach end of an instrument runway. When signal is received, the horizontal needle of the cross-pointer indicator shows whether the aircraft is above, below, or on the glide-slope.

An automatic landing system will help the aircraft to land without visual reference to the runway. There are a number of methods for automatic landing approach. If an aircraft employs a GPS, it may be guided along a geometrically defined line toward the runway, without a need to localizer and glide slope receivers.

4.3   APPROACH LOCALIZER TRACKING

From the beginning of the initial approach to the flare segment, or to a point from which a landing may be made visually, a series of maneuvers are needed for an aircraft. In airports with ILS, the incoming aircraft should detect airport localizer (i.e., horizontal component of ILS) beam, and adjust its heading with the localizer signal. The airport ILS broadcasts signals [13] to arriving instrument aircraft to guide them to the runway. Localizer information is displayed on the same indicator as for VOR information.

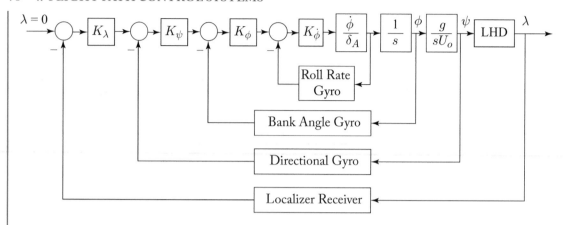

Figure 4.5: Block diagram of the localizer hold mode. LHD: Localizer Hold Dynamics.

The angle between the aircraft heading (Ψ in Figure 4.3a) and localizer beam should be less than 30°, before the auto-land can be turned on to follow the localizer. When autopilot for landing is on, the aircraft must catch the localizer, which implies that the localizer beam must be detected by aircraft receiver. AFCS is utilizing a radio beam directed upward from the ground to provide azimuth information, so that the aircraft can be lined up with the runway.

The approaching aircraft must be equipped with dual localizer and glide slope receivers to be able to receive two incoming beams. An important part of approach operation is governed by the localizer hold mode of the AFCS, which employs the heading command system and the horizontal guidance system. The approach localizer hold mode is under lateral-directional navigation function category of the AFCS.

The reference command is governed by the intercept beam and localizer dynamics. The hold mode will create a command until the localizer error angle is zero ($\lambda = 0$); and the aircraft is established on the localizer. The block diagram of the localizer hold mode has a number of subsystem; Figure 4.5 illustrates the main block diagram for landing approach localizer hold system.

To have the desired path and slope, all three control surfaces (aileron, elevator, and rudder) are frequently utilized. However, the primary control surface in the localizer hold mode is the aileron (δ_A). The rudder is employed as an interconnect to maintain a coordinated turn, while the elevator is deflected to adjust the pitch angle. For simplicity, the connections of these two control surfaces are not shown in the block diagram. The localizer hold mode incorporates the current aircraft heading angle (Ψ), and localizer error angle (λ) to create a reference heading angle (Ψ_{ref}).

The localizer error angle, λ, is determined from UAV distance (d) from the intended path (top-view) and slant range (R) as:

$$\lambda = 57.3 \frac{d}{R}. \tag{4.1}$$

The distance, d, in s-domain is determined from the integration of its rate of change

$$d(s) = \frac{1}{s}\dot{d}(s), \tag{4.2}$$

where

$$\dot{d} = U_o \sin(\psi - \psi_{ref}) \approx U_o(\psi - \psi_{ref}). \tag{4.3}$$

Therefore,

$$d(s) \approx \frac{U_o}{s} \left(\psi(s) - \psi_{ref}(s) \right). \tag{4.4}$$

This equation is employed to derive the localizer hold dynamics (e.g., transfer function). The task of the control system designer is to determine the values for four gains, K_λ, K_Ψ, K_ϕ, and $K_{\dot\phi}$ to satisfy the automatic landing requirements for localizer hold mode.

4.4   APPROACH GLIDE SLOPE TRACKING

After the localizer hold mode has corrected the heading of the aircraft, it is time to simultaneously correct (i.e., control) two flight parameters: flight path angle (γ) and height (h). The AFCS function is now to automatically descend according to the desired glideslope. In this operation, the pitch angle control system and airspeed control system should be active to allow a successful approach glide (path) slope hold mode. The glide slope capture cannot occur prior to localizer capture. The glide slope can be captured from above or below of the beam. When the glide slope annunciates is captured, the previous pitch mode should be disengaged. At the end of this phase, if the visibility is good, the runway should be in sight.

The approaching aircraft must be equipped with a glide slope receiver to be able to receive the incoming beam from airport transmitter (or a visual glide slope indicator). AFCS should employ a radio beam directed upward from the ground at about 2.5°. A commonly recommended approach glide slope is 2.5–3.5°. However, certain airports have steeper approach glide slopes due to on topography, buildings, mountain, or other location considerations. The glide path must usually be intercepted at, about 3,000-ft altitude, and the aircraft will descend with an airspeed of about 1.3 V_s.

The aircraft engine(s) during this approach phase is/are idle, so this flight phase can be called glide. The pitch control system is utilized to keep the aircraft on the desired glide slope (see Figure 4.4b). The primary control surface to control the flight path (in fact, the glide) slop is the elevator. The approach glide slope hold mode is under longitudinal navigation function category of the AFCS.

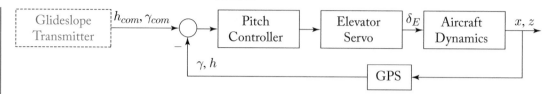

Figure 4.6: Block diagram of an approach path slope hold system when GPS is available.

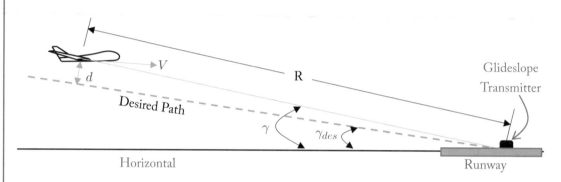

Figure 4.7: Airport transmitters transmit two guidance beams to approaching aircraft.

If the aircraft is equipped with an angle of attack (α) sensor, a control system is employed very similar to the pitch attitude control system (Figure 3.2). If the aircraft is employing GPS and the GPS signal is available, GPS equipment will directly calculate (Figure 4.6) the aircraft glide angle (γ) from the aircraft coordinates.

However, if the aircraft is not equipped with a sensor to measure the angle of attack, we need to only rely on the pitch attitude gyro, which measures the pitch angle (θ). Figure 4.7 demonstrates the geometry of an aircraft, which approaches an airport glide slope transmitter with an airspeed of V and a glide slope of γ. The objective of the aircraft glide slope hold is to guide and control the aircraft to follow the desired flight path in its landing approach.

To correct the flight slope, the pitch angle (θ) and angle of attack (α) must be simultaneously controlled. Both of these two parameters can be controlled by deflecting the elevator. To construct the block diagram of the glide slope hold system, four transfer functions should be derived ($\frac{\Gamma(s)}{d(s)}$, $\frac{d(s)}{\gamma(s)}$, $\frac{\gamma(s)}{\theta(s)}$, and $\frac{\theta(s)}{\theta_{com}(s)}$).

The glide slope transmitter broadcasts a beam (i.e., reference input signal to aircraft) at an angle of γ_{des}. Using triangle geometry shown in Figure 4.7, two parameters of d and slant range (R) are related as:

$$\tan(\Gamma) = \frac{d}{R}, \qquad (4.5)$$

where Γ is the angular deviation or glide slope angle difference. Note that the angle Γ is the difference between the desired glide angle and the aircraft glide angle:

$$\Gamma = \gamma_{des} - \gamma. \tag{4.6}$$

If the AFCS equipment onboard the aircraft can measure both the angular deviation (Γ) from the airport transmitter beam, and the distance R, it will calculate the perpendicular displacement of the aircraft from the glide path (i.e., glide-path deviation, d) from Equation (4.5). To control the height, the instantaneous distance (d) between aircraft center of gravity to the desired glide path must go to zero. Thus, in other words, the reference command in the glide slope hold system is:

$$\Gamma_{ref} = 0.$$

Since this angle (Γ) is very small (less than 5°), the Equation (4.5) can be linearized as:

$$\Gamma = \frac{d}{R}. \tag{4.7}$$

Applying Laplace transform, we obtain:

$$\Gamma(s) = \frac{d(s)}{R}. \tag{4.8}$$

Therefore, the first transfer function will be obtained as:

$$\frac{\Gamma(s)}{d(s)} = \frac{1}{R}. \tag{4.9}$$

The velocity with which the aircraft approaches the desired approach path is a component of the aircraft airspeed (V) as:

$$\dot{d} = V \sin\left(\gamma_{des} \pm \gamma\right). \tag{4.10}$$

The aircraft should maneuver so that the glide-path deviation, d is driven to zero. The aircraft is considered to approach the runway with a constant airspeed of V. In Equation (4.9), the $\pm$ sign is used to cover two cases: (1) when the aircraft is above the desired path ($-$) and (2) when the aircraft is under the desired path ($+$). The angle $\gamma_{des} \pm \gamma$ is usually very small, so, Equation (4.10) is linearized too:

$$\dot{d} = V \cdot \left(\gamma_{des} \pm \gamma\right). \tag{4.11}$$

Applying Laplace transform, we obtain:

$$sd = V \cdot \left(\gamma_{des}(s) \pm \gamma(s)\right). \tag{4.12}$$

Thus, the second following transfer function is derived as:

$$\frac{d(s)}{\gamma(s)} = \frac{V}{s}. \tag{4.13}$$

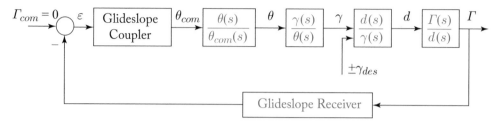

Figure 4.8: Block diagram of an approach path slope control system.

Please note that the term γ_{des} is removed in the transfer function; this term is added to the transfer function in the block diagram (as the second input).

Next, the glide-angle-to-pitch-angle transfer function—$\gamma(s)/\theta(s)$—is developed by using the relation between glide angle, pitch angle, and angle of attack ($\theta = \gamma + \alpha$). When we apply Laplace transform, and divide both sides of this relation by $\theta(s)$, we obtain:

$$\frac{\gamma(s)}{\theta(s)} = 1 - \frac{\alpha(s)/\delta_E(s)}{\theta(s)/\delta_E(s)} = 1 - \frac{\alpha(s)}{\delta_E(s)}\frac{\delta_E(s)}{\theta(s)}. \tag{4.14}$$

Two transfer functions ($\alpha(s)/\delta_E(s)$ and $\theta(s)/\delta_E(s)$) have already been presented in Chapter 3 as in Equations (3.3) and (3.9). After substituting both transfer functions and a manipulation, the following result is obtained:

$$\frac{\gamma(s)}{\theta(s)} = 1 - \frac{Z_{\delta_E}s + M_{\delta_E}(U_1 + Z_q) - M_q Z_{\delta_E}}{(M_{\delta_E}U_1 + Z_{\delta_E}M_{\dot\alpha})s + M_\alpha Z_{\delta_E} - Z_\alpha M_{\delta_E}}. \tag{4.15}$$

Finally, the transfer function $\frac{\theta(s)}{\theta_{com}(s)}$ is developed by combining all transfer functions involved in the block diagram of the pitch attitude control system, as already presented in Chapte 3 (Figure 3.2). As seen in Figure 4.8, the pitch attitude control system plays the role of the internal loop for the glide slope hold system. This inner loop includes a controller, which must be designed within the glide slope hold system design process. By implementing all derived transfer functions, the block diagram for glide slope hold system is established, as shown in Figure 4.8.

The function of glide slope coupler is to produce an input for the pitch attitude control system by processing the glide slope error signal ($\varepsilon = \Gamma_{com} - \Gamma$). Please note that the units of glide angles and pitch angle must be all in radian in their representative transfer functions. Moreover, the glide angle must have a negative value.

The challenging job of the control system designer is: to design the controller for the attitude control system and to design the glide slope coupler. A typical coupler has a proportional-integral (PI) transfer function.

The glide slope receiver is the measurement device for this control system to detect and measure the desired glide slope, and then to generate the glide slope angle error (Γ). When the aircraft glide angle is changed, the glide slope error is sensed by the aircraft receiver and fed back to the coupler. The glide slope receiver may be modeled with a pure gain or a first-order system.

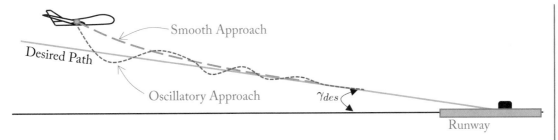

Figure 4.9: Two possible approach trajectories.

To improve the efficiency of the coupler, the aircraft may be equipped with a distance measuring equipment to measure the slant range (R). The coupler gain can be made as a function of the slant range, in order to make the coupler robust. This will make the approach flying qualities more acceptable with a smoother flight and a lower load factor. Another reason is to have a controllable touchdown point with an acceptable touchdown rate, which helps the landing gear to experience no damage.

Figure 4.9 illustrates two possible approach trajectories: smooth and oscillatory. The glide slope control system (including the controller) should be designed such that to help the aircraft to have a smooth approach (exponential). The goal is to keep up with the desired glide slope by increasing/decreasing the descent rate. Please note that before the touchdown the aircraft will flare to increase the angle of attack (in order to decrease the descent rate). The automatic flare mode is presented in Section 4.5.

4.5   AUTOMATIC FLARE CONTROL

The final phase of the landing approach is a transition from the glide slope to the touchdown, usually referred to as the flare. The landing flare is the transition phase between the final approach and the touchdown on the runway. After the AFCS has adjusted two flight parameters: the heading angle (localizer hold mode) and the glide slope angle (glide slope hold mode) for an approaching aircraft; it is the time for a flare operation which happens before the touchdown. A localizer hold mode has already made the aircraft to line up with the runway (initial approach phase), while the glide slope hold mode has captured the aircraft glide path along desired flight slope (intermediate approach phase). At the end of the glideslope tracking phase, another control system, the automatic flare control, is switched in. At the starting point of flare, if the visibility is good, the runway is in sight. For a transport aircraft, usually flare engages at 50 ft and retard engages at 27 ft.

If flare is not executed correctly, it could result in a hard landing, which may lead to: collapse of the landing gear; tail strike; or runway overrun. Moreover, a number of incidences that can happen—if the flare is not controlled correctly—include: excessive speed at the touchdown;

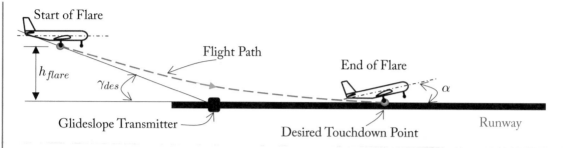

Figure 4.10: Geometry of flare.

excessive rate of descent at the touchdown; and overly aggressive pitch changes which could result in ballooning.

During flare, the landing gear is down and flap deflection is set for the landing conditions. During the flare (as the final approach phase), the nose of the aircraft is raised (pitching up), to slow the descent rate, and to set the proper attitude for touchdown. Hence, at the touchdown, an aircraft with a tricycle landing gear will have a high angle of attack. Figure 4.10 illustrates the geometry of flare, which includes the starting and end-points of flare and the typical flight path. The flare operation starts at a height of h_{flare} (or h_f) which is regulated by the FAA. Furthermore, the FAA has regulated that the distance from touchdown point to the glideslope transmitter (e.g., 1,100 ft for transport category aircraft). During the flare, the airspeed is assumed to remain constant, while the altitude decreases exponentially to zero.

The automatic flare control system makes the aircraft altitude go to zero, and touchdown with an acceptable rate of glide ($\dot{h}$). If the flare is not executed, the aircraft will hit the ground hard. In such undesirable situation, the decent rate is unacceptably high, which may hurt the passengers, aircraft structure, or landing gear. The recommended rate of descent at the touchdown is less than about 2 ft/sec by touchdown.

If the flare is too high, the aircraft may stall. However, this is not dangerous, because the aircraft should only be a couple inches off the runway, when it stalls. Hence, the aircraft will just settle down on to the runway.

An automatic flare control commands the aircraft to fly along an exponential path (see Figure 4.10) from the initiation of the flare until touchdown. An exponential curve with a negative slope has the following mathematical model:

$$h = h_o e^{-\frac{t}{\tau}}, \tag{4.16}$$

where h_o is the initial height (here, h_{flare}) and τ is the time constant. If the τ is known, the h_o can be determined, since at $t = 0$, $h = h_o$. By differentiating Equation (4.16), we obtain:

$$\dot{h} = -\frac{h_o}{\tau} e^{-\frac{t}{\tau}}. \tag{4.17}$$

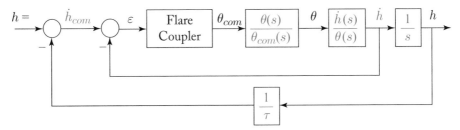

Figure 4.11: Block diagram of a flare control system.

At the start of the flare ($t = 0$),

$$\dot{h}_o = -\frac{h_o}{\tau} \Rightarrow h_o = -\tau \dot{h}_o. \tag{4.18}$$

On the other hand, the descent rate is a component of the airspeed as:

$$\dot{h} = V \sin(\gamma). \tag{4.19}$$

Due to small glide angle during flare, this equation can be linearized:

$$\dot{h} = V\gamma. \tag{4.20}$$

At the start of the flare, both airspeed and glide angle are known:

$$\dot{h}_o = V_{app} \sin(\gamma_{des}). \tag{4.21}$$

The desired glide angle at the start of flare is recommended to be 2.5°, and for each aircraft, the approach speed is known. Thus, the descent rate at the beginning of a flare is determined. The desired value of τ can be obtained by specifying the distance to the touchdown point from the glide slope transmitter. This distance is recommended to be about 1,000 ft. The typical value for the flare time constant (τ) is about 1–2 sec. During the flare, pitch angle (θ) and thus the rate of descent, $\dot{h}$, will be controlled by the pitch attitude control system, which is discussed in Chapter 3.

There are various methods to control the flare; each one requires a different control structure. Figure 4.11 shows a typical block diagram of the automatic flare system, which employs the pitch attitude control system as the **inner loop**. In the following, the transfer functions and controllers of this block diagram are discussed.

In the block diagram of Figure 4.11, there are three basic transfer functions ($\frac{\theta(s)}{\theta_{com}(s)}$, $\frac{\dot{h}(s)}{\theta(s)}$, and $\frac{1}{\tau}$) and three controllers. The transfer function, $\frac{\theta(s)}{\theta_{com}(s)}$, is developed by combining all transfer functions involved in the block diagram of the pitch attitude control system, as already presented in Chapter 3 (Figure 3.2). The $\frac{1}{\tau}$ is the control law at the outer feedback loop, as involved

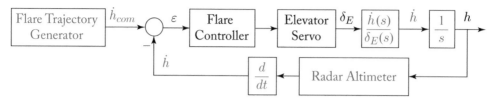

Figure 4.12: Block diagram of a flare control system using a radar altimeter.

in Equation (4.16). To develop the transfer function $\frac{\dot{h}(s)}{\theta(s)}$, the Laplace transform is applied to Equation (4.20):

$$\dot{h}(s) = \gamma(s)V, \tag{4.22}$$

where it is assumed that the airspeed (V) remains constant during flare. By dividing both side by $\theta(s)$, we obtain:

$$\frac{\dot{h}(s)}{\theta(s)} = \frac{\gamma(s)}{\theta(s)}V. \tag{4.23}$$

Please note that the unit of pitch angle and descent angle should be in radian. Moreover, both pitch angle and glide angle have negative values. The transfer function $\frac{\gamma(s)}{\theta(s)}$ has already been derived in Equation (4.15) and repeated here for convenience:

$$\frac{\gamma(s)}{\theta(s)} = 1 - \frac{Z_{\delta_E}s + M_{\delta_E}(U_1 + Z_q) - M_q Z_{\delta_E}}{(M_{\delta_E}U_1 + Z_{\delta_E}M_{\dot{\alpha}})s + M_\alpha Z_{\delta_E} - Z_\alpha M_{\delta_E}}. \tag{4.24}$$

The challenging job of the control system designer is: to design the controller for the attitude control system and to design the flare coupler. The function of flare coupler is to produce an input for the pitch attitude control system (θ_{com}) by processing the altitude error signal ($\varepsilon = h_{com} - h$). A typical coupler has a proportional-integral (PI) transfer function. The aircraft longitudinal dynamic stability during flare is strongly influenced by the gain of the coupler gain. Adding a lead network (i.e., lead-lag compensator) to the original coupler transfer function will lead in a higher value of coupler sensitivity, thus preventing the aircraft from touching down too soon. The final design goal is to produce a satisfactory flare flying qualities by having an acceptable descent rate ($\dot{h}$).

If the aircraft is equipped with a radar altimeter, there is an easier technique for automatic flare control system. Figure 4.12 shows the block diagram of a flare control system using a radar altimeter. The controlled variable is the descent rate ($\dot{h}$) which can be derived from the output of the radar altimeter. The commanded signal ($\dot{h}_{com}$) is provided by a flare trajectory generator, and the AFCS has to be able to efficiently track this command.

The flare representative aircraft longitudinal dynamics (i.e., transfer function $\frac{\dot{h}(s)}{\delta_E(s)}$) is derived through dividing both numerator and denominator of Equation (4.20) by elevator deflection (δ_E):

$$\frac{\dot{h}(s)}{\delta_E(s)} = V \frac{\gamma(s)}{\delta_E(s)}, \tag{4.25}$$

where the airspeed (V) is assumed to remain constant during flare. Since the pitch angle (θ) is equal to the sum of the angle of attack (α) and climb angle (γ), we obtain:

$$\frac{\dot{h}(s)}{\delta_E(s)} = V \left[\frac{\theta(s)}{\delta_E(s)} - \frac{\alpha(s)}{\delta_E(s)} \right]. \tag{4.26}$$

Both pitch-angle-to-elevator-deflection transfer function ($\frac{\theta(s)}{\delta_E(s)}$) and angle-of-attack-to-elevator-deflection transfer function ($\frac{\alpha(s)}{\delta_E(s)}$) have already been introduced in Chapter 3 (Section 3.4) as Equations (3.3) and (3.9). They are repeated here for convenience:

$$\frac{\theta(s)}{\delta_E(s)} = \frac{\left(M_{\delta_E} U_1 + Z_{\delta_E} M_{\dot{\alpha}} \right) s + M_\alpha Z_{\delta_E} - Z_\alpha M_{\delta_E}}{s U_1 \left\{ s^2 - \left(M_q + \frac{Z_\alpha}{U_1} + M_{\dot{\alpha}} \right) s + \left(\frac{Z_\alpha M_q}{U_1} - M_\alpha \right) \right\}} \tag{4.27}$$

$$\frac{\alpha(s)}{\delta_E(s)} = \frac{Z_{\delta_E} s + M_{\delta_E}(U_1 + Z_q) - M_q Z_{\delta_E}}{U_1 \left\{ s^2 - \left(M_q + \frac{Z_\alpha}{U_1} + M_{\dot{\alpha}} \right) s + \left(\frac{Z_\alpha M_q}{U_1} - M_\alpha \right) \right\}}. \tag{4.28}$$

The flare trajectory generator has five known values: (1) initial descent rate ($\dot{h}_o$); (2) descent rate at the touchdown ($\dot{h}_{td}$); (3) initial flare height (h_f); (4) final height at touchdown ($h_{td} = 0$); and (5) down range. Among many candidates for desired descending path, two acceptable flare paths that can be mathematically generated (as a function of distance, x) are: exponential and cubic polynomial:

$$\dot{h} = ax^3 + bx^2 + cx + d \tag{4.29}$$

$$\dot{h} = \dot{h}_f e^{-k_1 x}. \tag{4.30}$$

The a, b, c, d, and k_1 are constants to be determined from desired initial values of altitudes, descent rates, and down range. For a transport aircraft with an initial descent rate ($\dot{h}_f$) of 10 ft/sec, and a final descent rate of 2 ft/sec, the value of k_1 is about 0.0016 and d is 50 ft/sec. To determine the exponential flare trajectory, we need to have an initial value (h_f), and a time constant for an exponential equation. An exponential variations of the height (h) as a function of distance (x) will be:

$$h(x) = h_f e^{-k_2 x}. \tag{4.31}$$

For a transport aircraft with an initial height (h_f) of 50 ft, and a down rage (x) of 1,000 ft, the k_2 is about 0.00(4) Figure 4.13 shows typical acceptable variations of descent rate during flare, and Figure 4.14 illustrates a typical flare trajectory.

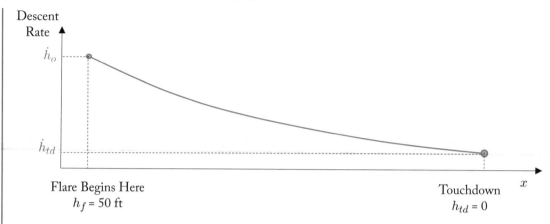

Figure 4.13: Typical variations of descent rate during flare.

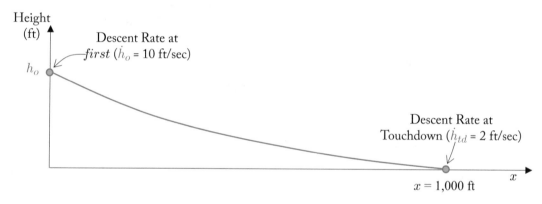

Figure 4.14: Typical flare trajectory.

The trajectory generation approach at approach and landing can be based [16] on the motion primitives (MPs) scheme, which consists of trims and maneuvers. This trajectory generation involves transitioning from one class of MPs to the other. All feasible trajectories should satisfy certain constraints such as optimality.

The objective is to produce a satisfactory flare flying qualities by having an acceptable descent rate ($\dot{h}$) at the moment of touchdown. The flare controller should be designed such that the maximum required elevator deflection is within the allowable values. The controllers are specific for the flare and touchdown of an automatic landing system. To deal with a turbulent air, a relatively low gain for the controller, as well as using an appropriate filter is recommended.

If an aircraft touches down at a descent rate of from 0–3 ft/sec, it is referred to as the soft landing, and it is highly recommended. However, if the descent rate at the touchdown is from 3–12 ft/sec, it is considered as a hard landing. This range of descent rate at the touchdown is not acceptable from flying qualities point of view (vertical speed in excess of limits). Once the main landing gear is in contact with the runway, de-rotation should occur without delay.

The above discussion did not consider the ground effect, that would be present in an actual landing the ground effect will generally reduce the rate of descent, which naturally improve the flare performance. The interested reader is recommended to include the ground effect in the design of the flare control system.

If there is a crosswind during landing, automatic landing system should yaw the aircraft while lining up the runway, de-crab the aircraft, and then, makes the wings level immediately before touchdown. These operations will make the flare control complex. Provision of such is necessary, since there are frequent crosswinds during landing in low temperature seasons.

4.6   AUTOMATIC LANDING SYSTEM

After a long cruising flight, an aircraft has to descend to the destination airport. At the end of the descent, the aircraft is initializing the landing operation, which begins with an approach. On a typical flight, the landing phase accounts for approximately 1–2% of the total time. However, controlling the trajectory of an aircraft in landing is the most challenging task for a human pilot, particularly, in conditions such as bad weather or limited visibility. The air/space plane X-37—with a length of 8.9 m, and a launch mass of around 5,000 kg—is equipped with a control system to land automatically upon returning from orbit.

Automatic landing should work in zero vertical visibility and very low horizontal visibility. An autopilot is very suitable equipment to tackle such risky operation. In a conventional fixed-wing aircraft, automatic control of the longitudinal trajectory requires simultaneous control of engine thrust and elevator. During landing, the landing gear is down and flap deflection is set for the landing conditions.

At an altitude of flare height (about 50 ft for transport category aircraft, and 35 ft for GA aircraft) above the runway, the AFCS must start to adjust the heading angle of the aircraft, achieve the correct pitch attitude for landing, and begin to reduce the rate of descent and airspeed further.

A landing is typically divided into four (Figure 4.15) main segments: (1) initial approach; (2) middle section of the approach; (3) flare as the last part of the approach; and (4) deceleration and ground roll. Although in an automatic landing, all flight parameters are simultaneously controlled, but it is recommended to focus on one flight variable at a time (i.e., each lading segment). Table 4.2 presents the main controlling variables in an automatic landing at various segments.

There are basically two methods for automatic landing: (1) using airport ILS and (2) smart tracking using GPS. In the first case, the aircraft receives two signals form two airport transmitters: localizer and glide slope transmitter.

In the second case, the aircraft will have the satellite-assisted navigation, which will offer [14] a number of advantages including a higher accuracy. Satellite-based navigation systems

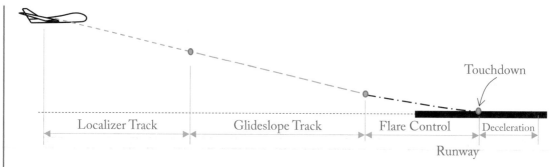

Figure 4.15: Four segments for AFCS modes in an automatic landing.

Table 4.2: Main controlling variables in an automatic landing

No	Landing Segment	Typical Length	Primary Varying Parameter	Remarks
1	Initial approach	6.5 nm	Heading angle (ψ)	Localizer (VHF) track, airborne
2	Approach middle part	3,500 ft	Glide angle (γ)	Glideslope (UHF) track, airborne
3	Flare	1,000 ft	Descent rate ($\dot{h}$)	High AoA at touch-down, airborne
4	Deceleration	Varies	Velocity (V)	Ground roll

are becoming a major element of navigation infrastructure. The system employs GPS signals for determining the aircraft coordinates (x, y, and z) in 3D space. Using GPS for landing requires installing the proper equipment at airports, equipping aircraft with additional devices and modifying new algorithms in AFCSs.

GA and regional business aircraft with simple panels usually are not equipped with the auto-land function; it is mainly available in large transport aircraft, which is assigned to fly long distances. Without a full auto-land capability, a pilot may not land if he/she cannot see the runway by 200 ft AGL.

The classic ILS does not provide landing of aircraft in all weather conditions. In aircraft using ILS, aircraft velocity is typically calculated by differentiating of position, or by integrating the acceleration outputs of an Inertial Measurement Unit. However, a GPS receiver directly measures three-dimensional velocity (ground speed) which is a very precise positioning sensor and is potentially available worldwide. This new measurement device is highly beneficial in de-

Table 4.3: Precision approach and landing categories

Category	Decision height (ft), (m)	Runway visual range (ft), (m)
I	200 ft (60 m)	2,400 ft (800 m)
II	100 ft (30 m)	1,200 ft (400 m)
III-A	100 ft (30 m)	1,200 ft (400 m)
III-B	50 ft (15 m)	170 ft (50 m)
III-C	No DA	No Runway Visual Range limitation

signing an automatic landing system. After the touchdown, the pilot needs to disengage the autopilot, maintain the runway centerline and decelerate manually.

To augment the GPS, the FAA has developed the **Wide Area Augmentation System** (WAAS) as an air navigation aid.

According to the FAA, as of July of 2018, more than 90,000 general aviation aircraft are equipped to fly WAAS-enabled procedures. Furthermore, about 4,000 Wide Area Augmentation System (WAAS) localizer performance with vertical guidance approach procedures serve close to 2,000 airports.

The FAA is working on Next Generation Air Transportation System (NextGen) to transition to performance-based navigation, which will result in many VOR sites that may be shut down. Aircraft equipped with WAAS can fly precise satellite-enabled RNAV approach procedures with localizers.

The satellite-enabled procedures—enabled by GPS with the Wide Area Augmentation System (WAAS)—is a network of ground stations and geostationary satellites that greatly enhances the accuracy, integrity, and availability of the GPS signal.

Three categories of landing (I, II, III-A, III-B, III-C) are defined according to decision height, visibility constraints and runway visual range (see Table 4.3). FAA has regulated the Air Traffic Control controllers how to handle these different kinds of approaches and visibility operations. For instance, CAT III-C is an actual zero visibility landing, and is regulated as "Not Authorized" by human pilot, unless an AFCS equipped with GPS receivers is employed.

Figure 4.16 shows the block diagram of the automatic landing control system. Its configuration includes four loops, three control surfaces, an engine throttle, and four controllers. Mainly, four flight parameters are measured by various aircraft measurement devices, and two beams (localizer and glideslope transmitters) are detected by aircraft receivers for tracking. Compared with the pitch angle (θ), the pitch rate (q) is a relatively faster state. Hence, in this configuration, the pitch rate feedback is an inner loop to the pitch angle feedback loop. Both of these flight parameters are controller via elevator and two controllers of C_1 and C_2.

The heading angle is controlled via both rudder and aileron. The ailerons are engaged through interconnect system (discussed in Chapter 3) to deliver a coordinated turn in changing

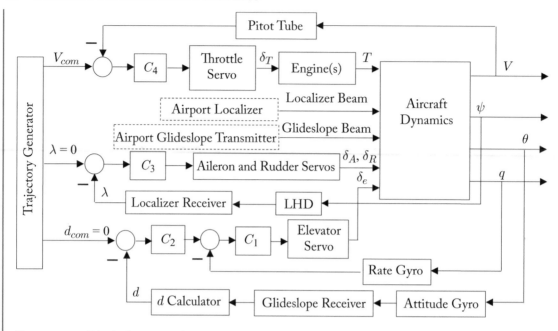

Figure 4.16: Block diagram of automatic landing system.

the heading. The controller C_3 is designed to control the heading angle to satisfy the lateral-directional flying qualities. The airspeed (in fact ground speed) is mainly controlled by engine throttle. The controller C_4 is designed to control the velocity and descent rate to satisfy the longitudinal flying qualities. In controlling velocity, pitch angle, and pitch rate, the aircraft angle of attack is varied, substantially in flare operation.

The design of controllers (C_1, C_2, C_3, and C_4) for automatic landing system consists of solving two problems: trajectory generation problem and asymptotic tracking problem. Many designee requirements—including acceptable control performance, stability margins, robustness, low sensitivity to turbulence, and effectiveness—are usually taken into account in the design process.

The transport aircraft Boeing 737 Max (Figure 4.17) with a wing area of 127 m² and a maximum takeoff mass of 80,286 kg is equipped with an AFCS. Its flight tests—including automatic landing [15], avionics, and environment control system testing—were scheduled to run through 2017.

The aircraft is the fastest-selling in Boeing's history. More than 4,500 B-737 Max aircraft have been ordered by 100 different airlines globally. The Boeing 737 was designed as a short-range narrow body airliner, and first flew in 1967. There have been more than a dozen subsequent models and it remains the best-selling commercial aircraft ever made, with some 10,000 manufactured.

Figure 4.17: Boeing 737 MAX 7.

4.7   VOR TRACKING

Another navigation reference, which was invented and deployed after the ILS, is the VOR system. VOR is a type of radio navigation system for aircraft. It is made up of a ground station
(transmitter) and an aircraft receiver unit. It enables an aircraft to determine its position, magnetic bearing, and the distance to a fixed ground station. Since both VOR and localizer have a
similar function during landing, they are generally coupled in the cockpit mode control panel as
one knob (e.g., VOR/LOC).

The VOR ground station broadcasts a reference VHF radio signal and data, which allows
the aircraft receiver to derive its magnetic bearing (direction from the VOR station in relation
to Magnetic North) from the station. This line of position is referred to as the radial from VOR.
There are 360 radials; each radial is equal to one degree. The VOR station sends a distinct signal
along each of its 360 radials. When an aircraft is on the 360 radial, it means that the aircraft is
on the north of the VOR station. An aircraft is on the 090 radial, is on the east of the VOR.
Since pilots are relying on compass for directional guidance, the radials transmitted by VOR are
aligned with magnetic north. Hence, north on a VOR is Magnetic North. Each VOR station

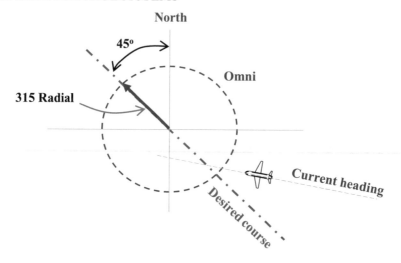

Figure 4.18: An example of VOR omni, desired course, and heading.

has a unique frequency and unique three-letter code identifire. So, in order for a pilot to receive the signal, he/she must dial the VOR frequency into the aircraft navigation receiver.

Figure 4.18 exhibits an example of VOR omni (315°), and an aircraft heading, when located southeast of the VOR. The desired course is along 315° radial, so this aircraft must first move to the left to intercept the radial; and then, turn right to track to the course. When this aircraft is on the 315° radial, and there is no crosswind, it will be heading toward the VOR.

The pilot should adjust the aircraft heading to make the track agree with the desired course. The pilot will use this information to determine the position and navigate the aircraft to a destination. When tracking a VOR, the pilot turns toward the needle in the same manner as with localizer navigation. This system can be used in both cruising flight and landing.

An aircraft can follow a cruising flight path from station to station by tuning into the successive VOR stations. These are traditionally used as intersections along airways in straight lines. Once a VOR station is identified, and the desired radial selected, one can select the navigation mode (the desired course on a VOR indicator) to track the selected radial.

A course deviation indicator is a measurement device to determine an aircraft lateral position (in x-y plane) from a radio navigation beacon (e.g., VOR ground station). The lateral deviation should indicate the distance from the selected radial, regardless of how close the aircraft is to the VOR ground station. Recall that the direction of the aircraft's nose is heading, while course is the desired track along the ground. Only when there is no crosswind, heading and course will be the same.

In many airports, both VOR and localizer are present to provide the directional navigation data. In these cases, a pilot has two options: use VOR or use localizer portion of an ILS. In the cockpit, the VORs share some equipment with the localizers such as: antenna, receiver, and some

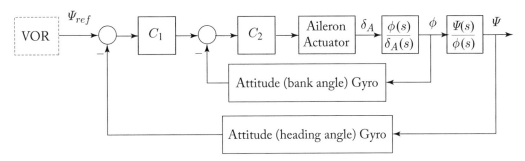

Figure 4.19: Block diagram of a VOR tracking system.

indicators. When pilot chooses a VOR station, the OBS will be functional and allows him/her to select the desired radial. When a localizer is chosen, the OBS is not functional, and the indicator is driven by a localizer converter. A VOR indicator will always show the actual radial from the Omni that the aircraft is on (regardless of the aircraft heading). Two main differences between VOR and localizer signals are: (1) width of the beam coming from the VOR station is smaller and (2) slant range will be larger in the VOR.

The frequencies of the VOR are in the range of 108.00–117.95 MHz (with 50-kHz spacing), and is different from the frequency range of a localizer. The signals are line of sight between ground transmitter and aircraft receiver and are useful for up to 200 miles. To leave channels for ILS, the 108.00, 108.05, 108.20, 108.25, and so on, are VOR frequencies, but 108.10, 108.15, 108.30, 108.35, and so on, are reserved for ILS in the U.S.

As of 2019, pilots still use VORs as a primary navigational aid. However, the VOR system is at the risk of being decommissioned by the FAA, due to the popularity of new technologies such as GPS, and automatic dependent surveillance-broadcast systems (ADS-B). Satellite-based navigational systems are increasingly replacing VOR, due to a lower overall cost and higher navigational performance. The FAA plans by 2020 to decommission roughly half of the 967 VOR stations in the U.S.

Whether the control objective of VOR tracking is: a cruising flight or landing; the block diagram is similar. A VOR tracking system is kind of an aircraft heading (indeed, magnetic bearing) control system, as shown in Figure 4.19. The bank angle control is employed as the inner loop, while the heading angle is the outer loop. This is due to the fact that aileron and rudder are connected as in an interlock to provide a coordinated turn. A vertical gyro is used for the purpose of measuring bank angle, and a directional gyro is used for the heading reference.

The bank-angle-to-aileron-deflection transfer function ($\frac{\phi(s)}{\delta_A(s)}$) is introduced in Chapter 3, and is repeated here for convenience. The approximate for the bank-angle-to-aileron-deflection transfer function is modeled by a second-order system as:

$$\frac{\phi(s)}{\delta_A(s)} = \frac{L_{\delta A}}{s^2 - sL_p}. \tag{4.32}$$

The $\frac{\Psi(s)}{\phi(s)}$ transfer function is:

$$\frac{\Psi(s)}{\phi(s)} = \frac{g}{sV}. \tag{4.33}$$

The C_1 is the controller for the outer loop (heading angle), and the C_2 is the controller for the inner loop (bank angle). Two controllers are designed simultaneously to obtain the desired yaw performance and gain acceptable flying qualities.

This is one technique to track a VOR signal; there are other methods to control the magnetic bearing). It is left to the interested reader to explore other methods. For instance, turn to a new heading could be accomplished by employing a turn control that will disengage the heading reference and command a yaw rate. When the turn is completed, the commanded yaw rate is changed to zero, and the heading reference will automatically reengage at the new heading angle.

In Figure 4.19, the effect of a crosswind—as a disturbance—is not included. To control the aircraft in the presence of a crosswind, the block diagram should be modified to include a loop for course deviation. In such a case, the reference heading (to be maintained) will be different from the commanded heading. Correcting for the wind drift requires a separate calculation and analysis.

4.8   AUTOMATIC FLIGHT LEVEL CHANGE

As discussed in Chapter 3, there are three ways to maintain a cruising flight: (1) constant-altitude, constant-angle-of-attack flight; (2) constant-airspeed, constant-angle-of-attack flight; and (3) constant-altitude, constant-airspeed flight. The option two is generally called a *cruise-climb*. However, due to safety concerns, most GA and transport aircraft are not allowed to gradually change their altitudes. Due to increasing air traffic and the assignment of distinct flight levels to specific flights, and directions of flight, it is no longer safe to climb continuously in this way.

The closest alternative is a flight operation called *step-climb*. Hence, most long flights compromise by climbing in distinct steps—a step climb—in order to ensure that the aircraft is always at an appropriate altitude for traffic control. For instance, in many long cruising flights, the cruise altitude (Figure 4.20) begins at 34,000 ft, and then, is gradually changed to 36,000 ft, 38,000 ft, and 40,000 ft. In returning flight, the cruise altitude begins at at 33,000 ft, and then, is gradually changed to 35,000 ft, 37,000 ft, and 39,000 ft. Almost every 2-3 hr, the flight altitude is increased 2,000 ft.

The AFCS is tasked to calculate a number of proper flight levels in a step climb, in order to maximize the efficiency. Thus, the objective of AFCS is to track the desired altitudes. An altitude tracking mode is employed to change the aircraft altitude from the current value to a new set value (e.g., cruising altitude). This mode generates guidance command to flight control system and auto-throttle to maintain constant airspeed while climbing or descending to a pre-selected altitude.

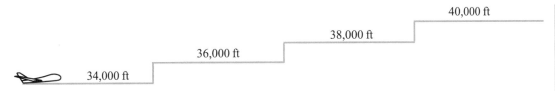

Figure 4.20: A typical step-climb.

This is a vertical flight profile and is frequently affected by speed and altitude restrictions specified in the published terminal procedures. Three related AFCS modes for altitude tracking are: Flight level change (FLC) mode; Vertical speed (VS) mode; and Vertical Navigation (VNAV) mode. These three modes can fulfill altitude tracking with various characteristics.

The FLC mode maintains airspeed during a climb or descent, while VS mode maintains a specific vertical speed. FLC is essentially airspeed-hold mode, and is typically only available in advanced autopilots. When FLC mode is active, selected altitude, airspeed, and current altitude should be continuously monitored. This mode acquires and maintains the reference airspeed while climbing or descending to the selected altitude. When the FLC mode is engaged during a climb/descent, the AFCS holds the aircraft in the climb or descent at the selected airspeed.

With the VNAV mode is engaged, the autopilot often commands a pitch, and employs the auto-throttle mode to fly the vertical profile selected on the display by pilot. The profile includes preselected climb, cruise altitude, speed, descent, and can also include altitude constraints at specified waypoints.

Figure 4.21 illustrates the block diagram of altitude tracking system for flight level change. The purpose of the altitude tracking mode is to change altitude at desired points during a cruising flight phase. At the instance of cruise altitude change, both elevator and throttle are utilized; however, during altitude hold function, the elevator suffices. The elevator is also employed to decrease the angle of attack as the aircraft weight is decreased. Two measurement devices of altimeter and airspeed meter are necessary for this autopilot mode.

The FLC controller could be as simple as a PID, or use a more advanced one. In either case, the aircraft will change the altitude and maintain the new altitude at each step. In designing the controller, one needs to use: (1) altitude-to-elevator-deflection transfer function and (2) velocity-to-throttle-setting transfer function. The derivation of these two transfer functions are presented in Chapter 3.

4.9   AUTOMATIC CLIMB AND DESCENT

After the takeoff, a transport aircraft will begin a climbing flight to a desired cruising altitude. At the end of cruising flight, the aircraft begins descending to the destination to land. These two climb and descent operation can be automated to reduce the pilot load, and to minimize the fuel cost. Descent phase begins at cruise altitude through approach to the beginning of the

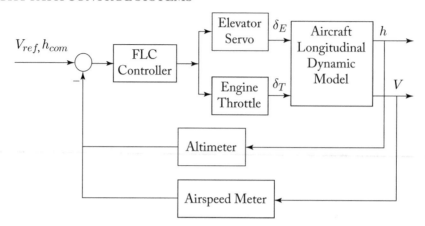

Figure 4.21: Block diagram of altitude tracking system.

missed approach (typically the runway). The approach may be assumed as a distinct phase or a segment within the descent phase.

The autopilot's vertical speed (VS) mode allows a pilot to perform constant-rate climb and descent. The speed-based climb is the operation, which employs the most economical airspeed [11]. In VS mode, the autopilot acquires and maintains a VS reference. In both climb and descent, the VS (climb rate and descent rate) is the vertical component (V_V) of the ground speed:

$$V_V = \dot{h} = V \sin(\gamma), \qquad (4.34)$$

where γ is the climb/descent angle. If there is no head-wind and no tail-wind, the ground speed is equal to the airspeed (V). A positive γ indicates that the aircraft is climbing, while a negative γ indicates that the aircraft is descending. In the case of a positive γ the VS is referred to as the rate of climb (ROC):

$$ROC = V \sin(\gamma) \qquad (\gamma > 0). \qquad (4.35)$$

In the case of a negative γ the VS is referred to as the rate of descent (ROD):

$$ROD = V \sin(\gamma) \qquad (\gamma < 0). \qquad (4.36)$$

As soon as an aircraft is over 500-ft altitude, one of the autopilot mode (including automatic climb) can be selected. For this goal, the pilot needs to set an altitude on the display to command the autopilot to climb to a cruising altitude. The VS mode gives pitch commands to hold the selected VS. With VS, the pitch holds the climb rate constant, and the throttles vary to hold the airspeed. The VS hold mode is employed in the climb routinely in modern transport aircraft. As an alternative, using FLC mode when climbing, will maintain the best airspeed (best climb rate).

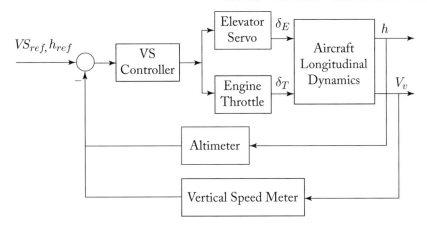

Figure 4.22: Block diagram of VS control system.

Figure 4.22 illustrates the block diagram of VS control system for climbing/descending flight. The purpose of this control mode is to maintain a desired (optimum) VS during a climb/descent phase. During climb/descent, both elevator and throttle are utilized. The elevator is employed to change both angle of attack and pitch angle, as the aircraft weight is decreased. Two measurement devices of altimeter and vertical airspeed meter are necessary for this autopilot mode. The throttle is utilized to maintain the optimum airspeed.

The vertical speed indicator (VSI) is one of the six basic flight instruments in aircraft to reflect the rate of climb or descent. This measurement device can be based on GPS data, or use the static port of a pitot tube. The VS controller could be as simple as a PID; or use a more advanced (e.g., LQR) one. In designing the controller, one needs to use: altitude-to-elevator-deflection transfer function or vertical-velocity-to-throttle-setting transfer function.

The VS is mainly a function of ground speed, engine throttle, aircraft weight, and drag:

$$V \sin(\gamma) = \frac{(T - D)V}{W}. \tag{4.37}$$

Using two transfer functions of $\gamma(s)/\delta_E(s)$ and $u(s)/T(s)$—which were introduced in Chapter 3—one can derive the vertical-velocity-to-throttle-setting transfer function.

The pilot must be very careful to specify an appropriate VS. The aircraft will fly itself into a stall if the pilot commands the autopilot to climb at a rate greater than the maximum rate of climb. Moreover, in a descending flight, descent rate should be such that aircraft does not exceed V_{NE} and to prevent an over-speed. Once the aircraft reaches the assigned altitude, the VS function should be automatically disengaged, and the altitude hold mode can be engaged.

It is recommended not to use VS mode when climbing, because aircraft may stall, if insufficient thrust/power is added for the requested climb rate. Instead, use the FLC mode which is much safer. However, it is recommended to use the VS mode when descending, because it gets

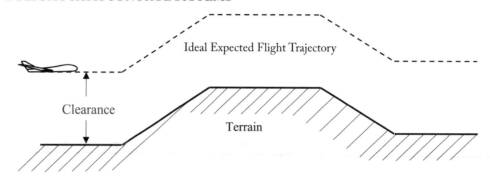

Figure 4.23: Terrain following.

down to the proper altitude in time. It also prevents physical discomfort for pilot and passengers. If the FLC is employed during the descent, be careful with thrust/power, in order not to exceed safe airspeed, and not to cause a discomfort (caused by excessive descent rates). Engine power will be adjusted by the autopilot to fly the aircraft at a pitch attitude corresponding to the reference vertical airspeed.

Another respective mode for automatic climb and descent is the VNAV mode which provides the vertical component of the flight plan (i.e., the computed flight trajectory of the airplane in the x-z plane). Vertical navigation provides flight control steering and thrust along the vertical path for takeoff, climb, cruise, descent, and approach. The VNAV and Lateral Navigation (LNAV) were first "fully integrated" on Boeing 757/767 airplanes in the early 1980s.

4.10   TERRAIN-FOLLOWING CONTROL SYSTEM

Some low-altitude applications of UAVs such as aerial photography/filming and agricultural/forest surveillance require the vehicle to follow the contour of the ground to capture high-resolution images. Moreover, some military applications of UAVs also concern terrain following (TF) flight operation to avoid enemy detection by flying as low as possible. In such applications, the objective is to ensure safety of the vehicle upon the ground, as well as the safety of the individuals in civic areas. Hence, the terrain scape maneuver should include a safe clearance (Figure 4.23) with respect to the ground.

The traditional terrain-following system is used for low-level flight by looking at the terrain ahead of the aircraft, and to providing the guidance command (through trajectory generation) to the flight control system to keep the flight at a specified altitude above the terrain. The vertical trajectory reference can be generated by a trajectory generating module using terrain elevation data. The primary goal in trajectory generation is to design a path that follows the terrain and avoids obstacles so that will satisfy the flight requirements, and aircraft constraints.

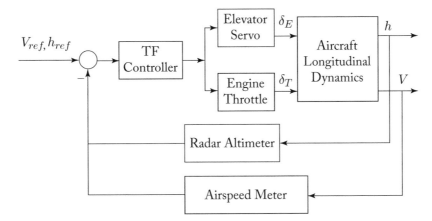

Figure 4.24: Block diagram of a terrain following control system.

Figure 4.24 illustrates the block diagram of a terrain following control system. During acceleration/deceleration and climb/descent, both elevator and engine throttle are utilized. The elevator is employed to adjust the altitude, while the throttle is varied to control the airspeed. Two measurement devices of radar altimeter and airspeed meter are necessary for this autopilot mode.

The TF controller could be as simple as a PID, or use a more advanced one (e.g., nonlinear, robust, optimal). For instance, Ref. [20] presents a minimum time trajectory planning in terrain following flights using the least square scheme to solve a general two-dimensional problem in the vertical plane.

Please note the difference between two related altitude terms: Height Above Ground Level (AGL) and Height above Mean Sea Level (MSL). For a cruising flight, the MSL is used through a conventional pressure altitude sensor (e.g., pitot-tube). However, in a terrain-following mission, the AGL should be utilized with sensors such as a radar altimeter or a lidar.

4.11   HEADING TRACKING SYSTEM

The heading mode is employed to automatically steer the aircraft along a pilot desired heading. In the beginning, an aircraft is cruising along a heading. However, the pilot wishes to change the heading to a new one (Figure 4.25) due to weather or mission destination. The task of the autopilot is to control the aircraft heading and to turn to an assigned heading safely until the new one is captured. The flight path under the autopilot control is shown with dashed lines.

Using the autopilot to fly a heading is done by: (1) selecting the assigned heading; (2) turning the knob to the new heading; and (3) engaging the heading function. After the autopilot captures the course, the heading tracking function automatically disengages, and the navigation function switches from "armed" to "engaged."

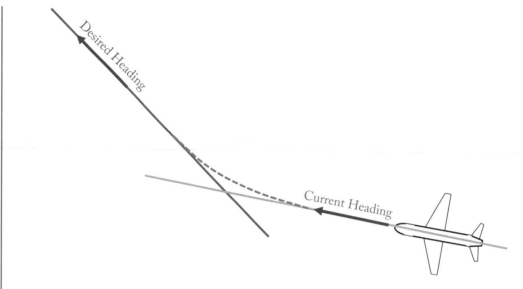

Figure 4.25: Heading change.

As stated in Section 4.7, a VOR tracking system is a kind of an aircraft heading control system. The general block diagram of a heading control system is very similar to the block diagram of a VOR tracking system as shown in Figure 4.19. The bank angle control is employed as the inner loop, while the heading angle is the outer loop. A vertical gyro is used for measuring bank angle, and a directional gyro is used for the heading reference.

A conventional autopilot mode in modern transport aircraft is the Lateral Navigation Mode (LNAV). In the LNAV mode, the autopilot controls roll/yaw to intercept and track a desired route. Since sufficient details for heading control/tracking system are provided in Section 4.7, they are not repeated here.

4.12   TRACKING A SERIES OF WAYPOINTS

In general, a waypoint is a term used to refer to an intermediate point or place on a route or line of travel, a stopping point or point at which course is changed. Waypoints are sets of coordinates that identify a point in physical space. The waypoints could be of two types: pre-defined and predicted (i.e., synthetic). In aviation, waypoints consist of a series of abstract GPS points (x, y, and z) that create artificial airways (i.e., highways in the sky). Thousands of waypoints can easily be programmed in an UAV software.

The path segments are frequently formed by altitude-constrained waypoints. The geometric path is a computed 3D point-to-point descent path between two constrained waypoints; or when tracking a prescribed vertical angle. The geometric path is a shallower descent and typically with a non-idle engine path.

Table 4.4: Typical approach waypoints

No	Waypoint	Approach Chart Text	Symbols in Charts
1	"AT" altitude	$\overline{\underline{4400}}$	⧖
2	"AT or Above"	$\underline{4400}$	△
3	"AT or Below"	$\overline{4400}$	▽
4	"Window"	$\overline{12,000}$ $\underline{4400}$	▽ △

The guidance system (based on an algorithm) determines a trajectory to follow in the path from waypoint 1 (or WP_n) to waypoint 2 (or WP_{n+1}). The algorithm should guide the vehicle along the assigned trajectory defined by a series of waypoints. A desired algorithm must have two functions: path planning and trajectory smoothing and tracking. When the aircraft is not on a waypoint, it determines the desired path from the current location of the vehicle to the desired waypoint.

When the waypoints are connected, a raw trajectory is generated. However, in many cases, such trajectory may not be feasible, considering the UAV constraints and limits. Airspeed, altitude, turn rate, and structural considerations limit maximum achievable acceleration level. Hence, computation of the flight path is influenced by factors such as aircraft weight, aircraft type and mission, wind (speed and direction), weather, temperature, turbulence. Table 4.4 shows typical approach waypoints as illustrated in the regularly published terminal procedures.

The vertical flight profile reflects the speed and altitude restrictions specified in the flight plan. Moreover, the lateral flight profile reflects the heading and airspace restrictions specified in the flight plan.

A combination of VNAV flight control and LNAV constitutes a reliable base for waypoint tracking. With the VNAV mode is engaged, the autopilot often commands a pitch, and employs the auto-throttle mode to fly the vertical profile selected on the display by pilot. The profile includes preselected climb, cruise altitude, speed, descent, and can also include altitude constraints at specified waypoints. Upon reaching the **last waypoint** in the VNAV flight plan, the autopilot may transition to altitude hold mode. Modern autopilots offer a global positioning system steering (GPSS) function. The GPSS performs typical navigation functions, but achieves a higher degree of precision due to the GPS receiver.

Various waypoints control techniques have been developed in the literature to minimize flight-path deviations. Such techniques [21] include applying control methods such as model predictive control and robust control to predict flight-path changes, and to take control action before reaching a waypoint. Reference [22] has implemented a waypoint-based guidance algorithm for a mini UAV via PID controllers for the outer loop navigation, and an adaptive controller for the inner loop variables.

4.13   DETECT AND AVOID SYSTEM

Collision avoidance is a primary concern [36] to the FAA regarding aircraft safety. There are many reports of civilian UAVs crashing into buildings, having hazardously close encounters with helicopters, peeping into residential windows, and being intentionally shot down. In a sense and avoid system, all three navigation, guidance, and control systems are working simultaneously. There is currently a large amount of research projects [37] being conducted in the area of detect (or sense) and avoid.

The standard for these airborne collision avoidance systems (ACAS) has been specified by International Civil Aviation Organization (ICAO). The traffic collision avoidance system (TCAS) is a particular ACAS implementation widely used in commercial aviation. However, at the present, there are no federal regulation concerning sense and avoid for unmanned aerial vehicle.

One of the major limitations to the widespread use of unmanned vehicles in civilian airspace has been the detect and avoid problem. The purpose of a sense and avoid system is to detect and resolve certain hazards to a UAV flight. The UAV detect/sense and avoid system must provide two services: self-separation service and collision avoidance service. In general, there are six functions required in a sense and avoid system: (1) detect the intruder/obstacle; (2) track; (3) evaluate; (4) calculation; (5) command; and (6) execute.

The function of the collision detection system is to detect a collision and provide the resolution in the form of an evasion maneuver. A collision is detected based on the flight path obtained from the trajectory prediction. The sense and avoid system checks the flight paths of both UAV and the other aircraft to evaluate whether the safety zone of the UAV has been intruded. If so, the location and times of the probable collision are calculated. Aircraft maneuverability has a direct impact on the reaction times during an avoidance maneuver or while planning a separation strategy.

In general, there are five collision avoidance functionalities for an air vehicle: (1) detect and avoid midair collisions with other flying traffic; (2) detect and avoid other flying objects (such as birds); (3) detect and avoid ground vehicles (when maneuvering on ground); (4) detect and avoid terrain and other obstacles (such as buildings or power lines); and (5) avoid hazardous weather (such as lightning).

Typical maneuvers to avoid a collision includes: left turn, right turn, climb, descent, slow down, speed up. These can be classified into three categories: turn, altitude change, and speed change. The collision avoidance resolution is derived from a suitable guidance law (e.g., Proportional Navigation, PN), and then executed by the control system of UAV's autopilot.

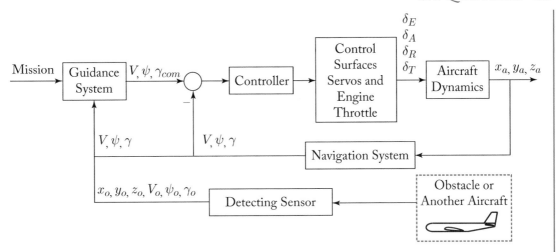

Figure 4.26: Block diagram of a basic sense and avoid system.

The block diagram of a basic detect and avoid system (Figure 4.26) has two main feedbacks: (1) flight parameters of the aircraft (x_a, y_a, z_a), and its velocity and directions (V, Ψ, γ); and (2) coordinates of the obstacle (x_o, y_o, z_o), and if it is moving, its velocity and directions (V_o, Ψ_o, γ_o). The guidance system will employ this information and the desired flight mission (i.e., reference trajectory) to generate the guidance commands. Detection sensors include measurement devices such as electronic camera (for day), infrared camera (for night), lidar, and radar. The coordinates of the vehicle are determined by navigation system at any instance.

The control system will create a control command based on the guidance command and the control law. All three control surfaces (e.g., elevator, aileron, and rudder) and the engine throttle are used to steer the aircraft to avoid any potential collision with the obstacle. Sadraey [21] presents the basic fundamentals of detect and avoid system. The interested reader is recommended to study this reference for the governing equations and the control laws.

4.14   QUESTIONS

4.1.   Name four functions provided by a closed-loop (i.e., negative feedback) control system.

4.2.   What three systems are working in conjunction with flight control system?

4.3.   What is another alternative name for the flight path control system?

4.4.   What is the primary function of a navigation system?

4.5.   List primary AFCS modes for longitudinal plane.

4.6.   List primary AFCS modes for lateral-directional plane.

4.7. What does VOR stand for?

4.8. Briefly describe the landing operation procedures.

4.9. What happens in the flare part of the landing?

4.10. What equipment/instruments are employed in ILS?

4.11. List three groups of information provided by an ILS to incoming aircraft.

4.12. What is a localizer?

4.13. What is the main function of the localizer hold mode?

4.14. What are the commonly recommended approach glide slope?

4.15. Draw the block diagram of an approach path slope hold system when GPS is available.

4.16. To correct the flight slope, which angles must be simultaneously controlled in glide slope tracking?

4.17. Briefly describe how a glide slope tracking system works.

4.18. What is the function of glide slope coupler?

4.19. What are the potential consequences, if the flare is not executed correctly (i.e., hard landing)?

4.20. Describe the flare operation.

4.21. What is the recommended rate of descent at the touchdown?

4.22. What is the desired glide angle at the start of flare?

4.23. Draw the block diagram of a flare control system.

4.24. Draw the block diagram of a flare control system using a radar altimeter.

4.25. Name four main segments of a landing operation.

4.26. List primary varying parameter for eachlanding segment.

4.27. What does WAAS stand for?

4.28. List precision approach and landing categories.

4.29. What sensors are employed in an aircraft automatic landing system?

4.30. What controls are employed in an aircraft automatic landing system?

4.31. Explain the differences between heading, radial, and course.

4.32. Briefly describe the differences between VOR and localizer and their signals.

4.33. Draw the block diagram of a VOR tracking system.

4.34. Describe three ways to maintain a cruising flight.

4.35. What are three related AFCS modes for altitude tracking?

4.36. Draw the block diagram of an altitude tracking system.

4.37. Describe the features of the autopilot's vertical speed (VS) mode.

4.38. Draw the block diagram of VS control system.

4.39. Draw the block diagram of a terrain following control system.

4.40. Explain how a terrain-following control system operate.

4.41. What is the difference between AGL and height above MSL?

4.42. Describe the waypoint.

4.43. What does GPSS stand for?

4.44. What does ACAS stand for?

4.45. What does ICAO stand for?

4.46. What does TCAS stand for?

4.47. List six functions that are required in a sense and avoid system.

4.48. List five collision avoidance functionalities for an air vehicle.

4.49. What are typical maneuvers to avoid a collision?

4.50. Draw the block diagram of a basic sense and avoid system.

CHAPTER 5

Stability Augmentation Systems

5.1   INTRODUCTION

The air through which an aircraft flies is dynamic, constantly in turbulent motion, and creates gusts. Consequently, the aerodynamic forces and moments fluctuate about their equilibrium values. These fluctuations will cause the aircraft to heave up or plunge down, to pitch its nose up/down, to roll about the x-axis, or to yaw from side to side about the aircraft's heading. These flight motions cause the large transport aircraft and high-performance fighters to suffer from low damping—particularly at low-speed and high-altitude flights. One solution is to augment the aircraft stability by employing AFCS. Without this solution, the damping ratio and natural frequency for open-loop dynamics of such aircraft deteriorate with altitude.

These motions result in accelerations, which are experienced by passengers and crew as unpleasant effects. To reduce these accelerations it is necessary to cancel the gust effects by other forces.

Modern high-performance commercial aircraft (e.g., Boeing 777 and Airbus 380) and military fighters (e.g., Lockheed MartinF-35 Lightning II) require some form of stability augmentation system. Current military fighter aircraft are actually dynamically unstable, and would be unsafe to fly without an AFCS and some form of stability augmentation system (SAS). SAS basically operate by sensing one or more of the aircraft motion variables, and then deflecting a control surface to oppose the aircraft motion. In general, SASs are concerned with the stabilizing/control of aircraft's motion in one axis. Feedback (usually electrical signals) from angular rate sensors are fed to the control surface actuators to modify the natural motion modes of the aircraft.

Automatic control devices for improving aircraft dynamic stability have been referred to—in the past—as stabilizers, dampers, and stability augmenters or stability augmentation systems. High-performance commercial (e.g., large transport) and fighter aircraft are required some form of stability augmentation system. As the name implies, a basic SAS will augment the static/dynamic stability of an aircraft to improve the aircraft response to atmospheric gust and disturbances. The role of SAS in large transport aircraft is mainly to increase the damping, while in fighters is to frequently stabilize the unstable aircraft. When AFCS is in SAS mode, it makes such aircraft appear to the pilot as normally responding aircraft.

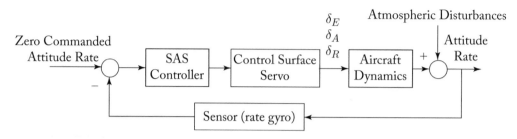

Figure 5.1: A general simplified block diagram of a stability augmentation system.

When an aircraft is unstable, or lightly damped, it is difficult for the pilot to control the aircraft. In terms of aircraft dynamics, the response modes determine if the aircraft is controllable. The responsiveness of an aircraft to maneuvering commands is partially governed by the frequency of the second order modes (e.g., short-period and Dutch roll). If frequencies of these modes are sufficiently high, the aircraft requires an SAS which artificially provides these modes suitable damping ratios and natural frequencies.

Short period mode of longitudinal dynamics and Dutch role of lateral-directional dynamics tend to deteriorate rapidly at low speeds, and high angles of attack.

When the SAS is switched on, the control surfaces are deflected by actuators which are controlled by the AFCS. When the SAS is switched off, the aircraft is controlled directly by the pilot by moving the appropriate control surface(s) through the cockpit stick/pedals.

Military fighters are lightly stable, or actually unstable, so they are virtually impossible to fly without an AFCS. To enhance the maneuverability of fighters, the longitudinal modes are intentionally made unstable, and directional modes are made lightly stable. Modern fighters are designed with a negative static margin in longitudinal dynamics.

Moreover, stability augmentation systems often form the inner loops of attitude control systems; which in turn, form the inner loops for the flight path control systems.

The SAS employs sensors to measure the aircraft body-axes angular velocities, and feeds back processed versions of these flight variables to controllers that drive the control surfaces. An SAS controller will generate an aerodynamic moment proportional to angular rate and its derivatives to cause a damping effect on the dynamics. Figure 5.1 demonstrates the general block diagram of an SAS. There could be one SAS for each axis; roll damper for x axis, a pitch damper for y axis, and a yaw damper for z axis.

Each SAS is equipped with a rate gyro to measure the relevant body-axis attitude (angular) rate. A roll rate gyro in a roll damper, a pitch rate gyro in a pitch damper, and a yaw rate gyro is employed in a yaw damper. If an aircraft dynamic mode (e.g., Dutch roll or short period) is unstable, or if it is desired to change both damping ratio and natural frequency (of the mode), additional feedback is required. When an atmospheric disturbance/gust hits the aircraft, the SAS will help the aircraft to respond as in a more stable aircraft.

If a flight mode is unstable, or if changes in both damping and natural frequency of a stable mode are independently desired, an additional feedback will be required. Stability augmentation systems are often less complex in their operation than attitude control systems, since typically less control surfaces are employed.

It is possible to use stability augmentation in conjunction with parallel systems. The SAS can be employed even when the pilot is controlling the aircraft with stick/pedal (i.e., manual control). However, the result is that SAS-induced control surface deflections are fed back to and felt by the pilot. Simultaneous control of an aircraft with both AFCS and human pilot is usually not acceptable. A notable exception is the yaw damper utilized in parallel with pilot controlled mechanical system found in many general aviation aircraft.

In flight conditions (e.g., cruising flight), where the aircraft longitudinal dynamics and lateral-directional dynamics are loosely coupled, two stability augmentation systems may be designed in parallel (one for longitudinal modes, and one for lateral-directional modes). In practice, longitudinal dynamics are decoupled from lateral-directional dynamics for non-maneuvering flight conditions.

In this chapter, six principal SAS functions are presented: (1) pitch damper; (2) yaw damper; (3) roll damper; (4) lateral-directional SAS; (5) stall avoidance system; and (6) automatic control of trim wheel.

5.2   PITCH DAMPER

Modern high-performance fighters suffer from low short period damping at low-speed and high-altitude flights. Moreover, most fighter aircraft are designed/built with inherent static longitudinal stability problems (e.g., statically longitudinally unstable). One solution to such problems is to augment the aircraft longitudinal stability by employing AFCS.

The pitch damper is a longitudinal control system that will use the elevator to further damp any disturbance in pitch beyond the output of natural longitudinal stability. This pitch control system is employed to augment the longitudinal stability characteristics of an aircraft. Thus, it benefits the horizontal tail design, when the pitch is lightly damped. The stability derivatives—which such system is aimed to improve—are Cm_q and Cm_α, which leads to an increase to the damping ratio of the short period mode.

In practice, an aircraft with a pitch damper will need a smaller horizontal tail; compared with an aircraft without a pitch damper. In this section, the stability augmentation application of an AFCS; pitch damper, is presented. The system ensures a good flying quality over the entire flight regime. Hence, an aircraft designer will achieve an improvement in longitudinal stability without compromising the level of longitudinal flying qualities. This will artificially increase the pitch damping ratios.

There are various techniques to augment longitudinal stability of an aircraft. A basic pitch damper will have only one feedback (pitch rate) and one sensor (pitch rate gyro). Another method is to measure the angle of attack or load factor by an appropriate measurement de-

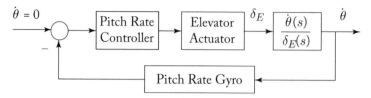

Figure 5.2: Block diagram of a basic pitch damper.

vice and provide feedback to the control system. A more effective technique is to have at least two feedbacks and two sensors to augment the aircraft longitudinal stability. These longitudinal stability augmentation techniques are presented in the following three subsections.

Another important function of a pitch damper is to avoid a pitch up (unstable flight condition). A pitch up—if not controlled—will end up a stall (high angle of attack), spin, and crash. An effective pitch damper will improve airworthiness of an aircraft.

5.2.1   BASIC PITCH DAMPER

A basic pitch damper is a longitudinal control system that will use the elevator to further damp any disturbance in pitch beyond the output of open-loop longitudinal stability. This pitch control system is employed to augment the longitudinal stability characteristics of an aircraft. Figure 5.2 demonstrates the block diagram of a basic pitch damper. The commanded and measured flight variable is the pitch rate. The actuator for this damping is the elevator, and the measurement device is the pitch rate gyro. Any disturbance in pitch angle (θ) will be damped, so the pitch angle will go to the set value.

The location of the pitch rate gyro must be chosen very carefully to avoid picking up the vibrations of the aircraft structure. The gyro filter is usually used to remove noise and cancel structural mode vibrations.

The pitch-rate-to-elevator-deflection transfer function has a fourth order characteristic equation. Expressions for the numerator and denominator coefficients can be derived by expanding the longitudinal state-space model. However, using the short period approximation, the following transfer function with a second-order characteristic equation [1] is obtained:

$$\frac{\dot{\theta}(s)}{\delta_E(s)} = \frac{\left[(U_1 - Z_{\dot{\alpha}})M_{\delta_E} + Z_{\delta_E}M_{\dot{\alpha}}\right]s + M_\alpha Z_{\delta_E} - Z_\alpha M_{\delta_E}}{U_1 \left\{ s^2 - \left(M_q + \frac{Z_\alpha}{U_1} + M_{\dot{\alpha}}\right)s + \left(\frac{Z_\alpha M_q}{U_1} - M_\alpha\right)\right\}}. \tag{5.1}$$

In a cruising/climbing flight, the commanded pitch rate ($\dot{\theta}$) is zero, while in a vertical maneuver (e.g., pull-up and push over), a non-zero pitch rate is commanded.

In a pull-up maneuver (i.e., the aircraft in the lower half of a vertical loop), the aircraft is turning vertically (pulling up) with a constant airspeed V, with a radius R. For such maneuver, the commanded (i.e., reference; $\dot{\theta}_{Ref}$) pitch rate is not zero and should be computed. The angular

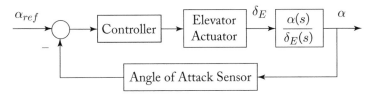

Figure 5.3: Block diagram of an angle of attack control system.

velocity (here, pitch rate) is defined as linear speed divided by radius, so the instantaneous turn rate [11] is:

$$\dot{\theta}_{Ref} = Q = \frac{V}{R} = \frac{g(n-1)}{V}, \tag{5.2}$$

where n is the load factor. Using gain scheduling technique, the controller gain may be scheduled to vary with altitude and airspeed. For instance, the gain at high altitudes could be a few times greater than that for sea level.

Pitch rate damper will oppose any pitch rate away for the reference command. If a faster pitch response is desired, another loop for an angle of attack feedback should be added.

5.2.2   STATIC LONGITUDINAL STABILITY AUGMENTATION SYSTEMS

Most fighter aircraft are designed/built with inherent static longitudinal stability problems (e.g., statically longitudinally unstable). One solution to such problems is to augment the aircraft longitudinal stability by employing angle of attack feedback. Figure 5.3 demonstrates the block diagram of an angle of attack control to augment the static longitudinal stability. Similar to a pitch rate damper, the elevator deflection is controlled to improve the aircraft response to gust and disturbance.

The measurement device is the angle of attack sensor that usually is typically of vane type. The angle of attack measurement may be obtained from the pitot-static air-data system, or a small "wind vane" mounted on the side of the aircraft forebody. The vane dynamics is always a concern, since it is too sensitive to the perturbations in the local flow. A solution to a sensitive angle of attack sensor is to employ a significant amount of output filtering. In addition, two sensors may be used, on opposite sides of the aircraft, to provide redundancy and possibly to average out measurement errors caused by side-slipping. The signal from the angle of attack sensor is usually noisy because of turbulence, and a filter is used to reduce the amount of noise injected.

The angle-of-attack-to-elevator-deflection transfer function has a fourth order characteristic equation. Using short period approximation, the angle-of-attack-to-elevator-deflection

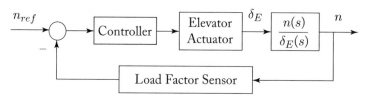

Figure 5.4: Block diagram of a load factor control.

transfer function is obtained [2] as:

$$\frac{\alpha(s)}{\delta_E(s)} = \frac{Z_{\delta_E} s + M_{\delta_E}(U_1 + Z_q) - M_q Z_{\delta_E}}{U_1 \left\{ s^2 - \left(M_q + \frac{Z_\alpha}{U_1} + M_{\dot\alpha} \right) s + \left(\frac{Z_\alpha M_q}{U_1} - M_\alpha \right) \right\}}. \tag{5.3}$$

Another longitudinal stability augmentation solution is to utilize a less sensitive measurement device; one example is a load factor sensor (Figure 5.4). The sensitivity of a load factor sensor to a change in the angle of attack (n_α) is:

$$n_\alpha = \frac{\partial n}{\partial \alpha} = \frac{\partial L/W}{\partial \alpha} = \frac{\partial \left(\frac{1}{2} \rho V^2 S C_L \right)/W}{\partial \alpha} = C_{L_\alpha} \frac{\frac{1}{2}\rho V^2}{W/S}. \tag{5.4}$$

The controller should be designed such that the longitudinal flying qualities are improved to an acceptable level (e.g., a short period frequency to about 2–3 rad/sec). The perturbed load factor can be written [2] in terms of climb angle as:

$$n = \frac{V\dot\gamma}{g}. \tag{5.5}$$

Thus, the corresponding load-factor-to-elevator-deflection transfer function will be:

$$\frac{n(s)}{\delta_E(s)} = \frac{V}{g} \frac{\dot\gamma(s)}{\delta_E(s)}. \tag{5.6}$$

Since the pitch angle is equal to the sum of the angle of attack and climb angle, we obtain:

$$\frac{n(s)}{\delta_E(s)} = \frac{V}{g} s \left[\frac{\theta(s)}{\delta_E(s)} - \frac{\alpha(s)}{\delta_E(s)} \right]. \tag{5.7}$$

Both pitch-angle-to-elevator-deflection transfer function ($\frac{\theta(s)}{\delta_E(s)}$) and angle-of-attack-to-elevator-deflection transfer function ($\frac{\alpha(s)}{\delta_E(s)}$) have been already introduced in Chapter 3 (Sec-

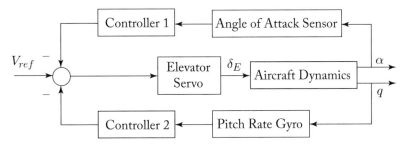

Figure 5.5: Block diagram of a pitch-axis stability augmentation system.

tion 3.4) as Equations (3.3) and (3.9). They are repeated here for convenience:

$$\frac{\theta(s)}{\delta_E(s)} = \frac{\left(M_{\delta_E}U_1 + Z_{\delta_E}M_{\dot{\alpha}}\right)s + M_\alpha Z_{\delta_E} - Z_\alpha M_{\delta_E}}{sU_1\left\{s^2 - \left(M_q + \frac{Z_\alpha}{U_1} + M_{\dot{\alpha}}\right)s + \left(\frac{Z_\alpha M_q}{U_1} - M_\alpha\right)\right\}} \tag{3.3}$$

$$\frac{\alpha(s)}{\delta_E(s)} = \frac{Z_{\delta_E}s + M_{\delta_E}(U_1 + Z_q) - M_q Z_{\delta_E}}{U_1\left\{s^2 - \left(M_q + \frac{Z_\alpha}{U_1} + M_{\dot{\alpha}}\right)s + \left(\frac{Z_\alpha M_q}{U_1} - M_\alpha\right)\right\}}. \tag{3.9}$$

The controller should be designed such that the required elevator deflection for the highest possible load factor perturbation (e.g., 6) is reasonable (e.g., 25°). To deal with a sensitive load factor sensor to a turbulent air, a relatively low gain for the controller, as well as using an appropriate filter is recommended.

5.2.3 PITCH-AXIS STABILITY AUGMENTATION SYSTEMS

Some air vehicles are provided with a pitch damper to artificially provide adequate damping in both phugoid and short period modes. When both frequency and damping of the short-period mode are unsatisfactory, or if the mode is unstable, two feedbacks are necessary (a pitch rate (q), and an angle of attack (α)). If an aircraft is required to perform a precise mission (e.g., air refueling), the pitch SAS will facilitate the mission. The main purpose of a pitch SAS is to provide desired natural frequency and damping ratio for the short-period mode of longitudinal dynamics.

This pitch SAS is employed to highly augment the longitudinal dynamic stability characteristics of an aircraft. Figure 5.5 demonstrates the block diagram of a pitch-axis stability augmentation system. The commanded flight variable could be such flight parameters as airspeed, pitch angle, pitch rate, or altitude. The actuator is the elevator, and two measurement devices are an angle of attack sensor and a pitch rate gyro. The phugoid mode will not be significantly affected by angle of attack feedback. For both feedback signals, filters (not shown in the figure) may be employed to filter the noise.

Figure 5.6: A U.S. Navy McDonnell Douglas F/A-18C Hornet (courtesy of Lt. Kyle "Chet" Turco). https://en.wikipedia.org/wiki/McDonnell_Douglas_F/A-18_Hornet#/media/File:FA-18-NAVY-Blue-Diamond.jpg

When the aircraft Cm_α is positive, the pitch SAS will make the slope of the pitching moment curve negative in the region around the operating angle of attack. Hence, the longitudinal static stability will artificially be maintained.

Two controllers (1 and 2) should be simultaneously designed such that the longitudinal modes behave so that the aircraft flying qualities in pitch are at an acceptable level. The goal of the angle of attack feedback is mainly to pull the unstable poles, from right hand side back into the left-half s-plane. This can be achieved by using root-locus design technique. In general, as the magnitude of the angle of attack feedback is increased, the frequency of the closed-loop short-period mode poles increases, and at the same time, they move toward the right-half plane.

The new position of the short period mode poles is recommended to be such that the augmented natural frequency is more than 1 rad/sec, and the augmented damping ratio is more than 0.3. The phugoid mode poles move very slightly with changing controllers gains. The pitch rate feedback mainly determines the damping ratio of the short-period mode. However, the angle of attack feedback usually stabilizes the unstable short-period mode and determines its natural frequency.

In the 1980s, the SASs were developed and applied to the longitudinal dynamics of five Navy [23] tactical aircraft: the A-6, A-7, S-3, F-14, and F-18 (Figure 5.6). The project was an attempt to utilize the existing requirements in analyzing advanced aircraft/control system configurations, by introducing the concept of equivalent systems.

The F-18 fighter airplane possesses [23] a highly complex digital flight control system. It incorporates numerous compensated feedbacks, stick shaping, lead-lag filters, and has separate control law configurations for both cruise and power approach flight conditions. The pitch rate response to control force inputs is described by a 14th order transfer function in cruise configuration, and by an eleventh-order transfer function in power approach.

5.3   YAW DAMPER

Modern high-performance fighters and large transport aircraft suffer from low Dutch roll damping at low-speed and high-altitude flights. In these cases, the yaw rate perturbations due to atmospheric turbulences can be very annoying to crew and passengers. One solution is to augment the aircraft directional stability by employing AFCS. The yaw damper can always be enabled during flight; it will not interfere with other AFCS modes. For instance, in 2014, Garmin conducted the maiden flight [27] of the Beechjet 400A featuring the G5000 Integrated Flight Deck. During the flight, the crew engaged and evaluated the autopilot and yaw damper.

The stability derivatives—which such system is aimed to improve, are Cn_r and Cn_β—which leads to an increase to the damping ratio of the Dutch roll mode. In this section, the stability augmentation application of an AFCS, yaw damper, is presented.

5.3.1   BASIC YAW DAMPER

A basic yaw damper is a directional control system that will use the rudder to further damp any disturbance in yaw beyond the output of natural directional stability. This yaw control system is employed to augment the directional stability characteristics of an aircraft.

The yaw damper is to augment the directional stability of an aircraft by avoiding/damping the yawing oscillation, known as Dutch roll. The purpose of the yaw-damper feedback is to use the rudder to generate a yawing moment that opposes any yaw rate that builds up from the Dutch roll mode. Most large aircraft with high-altitude cruising flight are equipped with such mode. The reason is that at high altitude, the turbulence is strong, and gusts hits the fuselage nose, and pushes the nose to the left and right almost continuously. The autopilot mode will keep the heading in the desired direction.

A yaw damper benefits the vertical tail design, when the yaw is lightly damped. In practice, an aircraft with a yaw damper will need a smaller vertical tail; compared with an aircraft without a yaw damper. The system ensures a good directional flying quality over the entire flight regime. Hence, an aircraft designer will achieve an improvement in directional stability without compromising the level of directional flying qualities.

Figure 5.7 depicts the block diagram of a basic yaw damper. The reference yaw rate is desired to be zero ($\dot{\psi} = r = 0$), so the yaw damper will drive any undesired yaw rate to zero. A rate gyro will measure the yaw rate, and provide a feedback for the system. The actuator for this damping is the rudder, and the measurement device is the yaw rate gyro; hence the sense output is the yaw rate. The function of a yaw damper is to oppose and eliminate any yaw rate perturbations which are caused due to atmospheric turbulences. The reference command is a zero yaw rate. Any disturbance in yaw angle (ψ) and yaw rate ($\dot{\Psi}$) will be damped, so the yaw angle and yaw rate will eventually go to zero. The controller will produce a controlling signal which is implemented through an actuator (i.e., rudder servo). The yaw damper controller can be as simple as a gain, K.

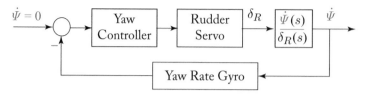

Figure 5.7: Block diagram of a basic yaw damper.

Many air vehicles are provided with a yaw rate damper to artificially provide adequate damping in Dutch roll (i.e., Dutch roll damper). When an aircraft is required to perform a precise mission (e.g., taking g picture), the yaw damper will facilitate the mission.

The approximate yawing moment equation as a function of stability and control dimensional derivatives [5] is:

$$\ddot{\psi} - N_r \dot{\psi} + N_\beta \Psi = N_{\delta R} \delta_R, \tag{5.8}$$

where N_r and N_β are dimensional yaw stability derivative, and $N_{\delta R}$ is the dimensional yaw control derivative. Applying the Laplace transform, this differential equation is readily converted to a transfer function:

$$\frac{\psi(s)}{\delta_R(s)} = \frac{N_{\delta R}}{s^2 - s N_r + N_\beta}. \tag{5.9}$$

By comparing this transfer function with the standard form for a standard second-order system (Equation (1.3)), one can conclude:

$$s^2 - s N_r + N_\beta = s^2 + 2\xi \omega_n s + \omega_n^2. \tag{5.10}$$

All corresponding terms in both sides should be equivalent, which results in the following:

$$-N_r = 2\xi \omega_n \tag{5.11}$$
$$N_\beta = \omega_n^2 \tag{5.12}$$

or

$$\omega_n = \sqrt{N_\beta} \tag{5.13}$$
$$\xi = \frac{-N_r}{2\omega_n} = \frac{-N_r}{2\sqrt{N_\beta}}. \tag{5.14}$$

A low damping ratio would result in a poor damping, and a large overshoot, which is very difficult for autopilot to control the flight direction in x-y plane. A reasonable design objective is to provide a damping ratio $0.35 < \xi < 1$, with a natural frequency $0.1 < \omega_n < 1$ rad/sec. These two objectives may be applied to Equations (5.13) and (5.14) to design a feedback control system such that the rudder deflection is proportional to the yaw rate:

$$\delta_R = -k\dot{\psi}. \tag{5.15}$$

Substituting the control deflection expression into Equation (5.8) yields:

$$\ddot{\psi} - N_r \dot{\Psi} + N_\beta \Psi = N_{\delta R} \left(-k \dot{\Psi} \right). \tag{5.16}$$

By rearranging, one can obtain:

$$\ddot{\psi} - (N_{\delta R} - k N_r) \dot{\Psi} + N_\beta \Psi = 0. \tag{5.17}$$

By proper section of factor k, one can satisfy the yaw damping requirement. So, it is concluded that a yaw damper is a AFCS device which artificially enhances the negative magnitude of the aircraft yaw damping derivative (N_r or $C n_r$).

A reasonable design objective is to provide a damping ratio greater than $0.3 (\xi > 0.3)$, with natural frequency less than one rad/sec ($\omega_n < 1$ rad/s). MathWorks [12] provides an example of a yaw damper design for a Boeing 747 Jet aircraft using Matlab.

A yaw damper may also be utilized during a turning maneuver. However, in a coordinated steady-state turn, the yaw rate has a constant nonzero value. When an aircraft tends to turn, and has a bank angle (ϕ_1), the yaw damper will try to fight with the constant bank angle turn. This is due to the fact that, the yaw rate (output of rate gyro) is a function of bank angle:

$$R_1 = \dot{\psi}_1 \cos(\theta_1) \cos(\phi_1), \tag{5.18}$$

where subscript 1 stands for steady-state value. One solution for such case is to add a first-order high-pass **washout filter** ($\frac{s\tau}{1+s\tau}$) with a time constant (τ) of about 4 sec in the feedback loop. This filter is applying a transient rate-feedback, in which the feedback signal is approximately differentiated. Although this filter causes this yaw rate to be gradually vanished, it will create a lag in the response of the yaw rate gyro.

A yaw damper is using the rudder to generate a yawing moment that opposes any yaw rate that builds up from the Dutch roll mode. However, in a coordinated turn, the yaw rate has a constant nonzero value. Hence, this raises a difficulty; where the yaw damper will try to oppose. Therefore, in a turn, with the yaw-damper operating, the pilot must apply larger than normal rudder inputs to overcome the operation of the yaw damper. One solution is not to use the yaw damper during a turning maneuver.

Another solution is to employ a first-order high-pass (washout) filter to implement an approximate differentiation. Due to a strong coupling between the roll and yaw channels, the washout filter time constant is determined in a compromise process: (1) a too large value is undesirable since the yaw damper will interfere with the entry into a turn; and (2) a too small value will reduce the achievable Dutch roll damping. An optimum gain value will ensure a safe turn with an acceptable roll performance (flying qualities).

5.3.2   SIDESLIP ANGLE FEEDBACK

Tailless fighter aircraft and aircraft with small vertical tail are designed/built with inherent static directional stability problems (e.g., statically directionally unstable). One solution to such prob-

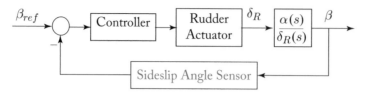

Figure 5.8: Block diagram of a sideslip angle control.

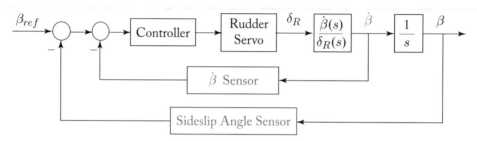

Figure 5.9: Block diagram of a $\beta - \dot\beta$ yaw SAS with two feedbacks.

lems is to augment the aircraft directional stability by employing sideslip angle feedback. Figure 5.8 demonstrates the block diagram of a sideslip angle control to augment the static directional stability. Similar to a yaw rate damper, the rudder deflection is controlled to improve the aircraft response to gust and disturbance in the x-y plane.

The measurement device is the sideslip angle sensor that is typically of a vane type. The vane dynamics is always a concern, since it is too sensitive to the perturbations in the local flow. It is difficult to accurately measure the direction of the local airflow, in determining whether the direction of the local airflow measured is the direction of the undisturbed air or the distorted local airflow. In spite of the difficulty in measurement of β the use of sideslip to achieve coordination is quite extensive. The reference sideslip angle (β_{ref}) is usually zero.

Another effective directional SAS is to control both sideslip angle (β) and its rate of change ($\dot\beta$). An inner-loop sideslip rate ($\dot\beta = d\beta/dt$) feedback is added as the second feedback to provide a better Dutch roll damping (i.e., Dutch roll damper). The new control system with two feedbacks, which requires another sensor, is illustrated in Figure 5.9. In this control system, the sideslip angle is measured with a regular sideslip angle sensor, the sideslip angle rate is measured a $\dot\beta$ sensor. There is an integrator ($1/s$) to derive the sideslip angle from sideslip angle rate in the outer loop. The reference sideslip angle (β_{ref}) is usually zero. A $\beta - \dot\beta$ SAS is superior to a regular yaw damper. This SAS was employed [4] in the design of the autopilot for the F-15 integrated flight/fire control system.

The sideslip-angle-to-rudder-deflection transfer function has a fourth order characteristic equation. Using Dutch roll approximation, the sideslip-angle-to-rudder-deflection transfer function is obtained [2] as:

$$\frac{\beta(s)}{\delta_R(s)} = \frac{Y_{\delta_R}s + N_{\delta_R}Y_r - N_{\delta_R}U_1 - N_rY_{\delta_R}}{s^2 - s\left(N_r + \frac{Y_\beta}{U_1}\right) + N_\beta + \frac{1}{U_1}\left(N_rY_\beta - Y_rN_\beta\right)}, \tag{5.19}$$

where U_1 is the initial trim airspeed, and other parameters are the stability and control lateral-directional derivatives.

The controller should be designed such that the Dutch roll mode behaves such that, the aircraft flying qualities in yaw are at an acceptable level. The minimum acceptable [10] Dutch roll un-damped natural frequency is 0.4 rad/sec.

5.4   ROLL DAMPER

Modern high-performance fighters, large transport aircraft, and most GA aircraft suffer from instability in lateral motions. When the bank angle is disturbed, such aircraft does not have a tendency to return back to its initial bank angle. One solution is to artificially stabilize the aircraft lateral dynamics by employing SAS. The stability derivatives—which such system is aimed to improve, are C_{l_p} and dihedral effect (i.e., C_{l_β})—which leads to an increase to the damping of the roll mode. In this section, the stability augmentation application of an AFCS in roll is presented.

The roll damper is a lateral control system that will use the aileron to further damp any disturbance in roll beyond the output of natural lateral stability (if any). This roll control system is employed to augment the lateral stability characteristics of an aircraft. Thus, it benefits the wing, horizontal and vertical tail design, when the roll is lightly damped. In practice, an aircraft with a roll damper will need a smaller wing, compared with an aircraft without a roll damper.

By convention, a positive deflection of the rudder generates a positive rolling moment and a negative yawing moment. The negative yawing moment rapidly leads to positive sideslip, which will in turn produce a negative rolling moment. These competing effects, tend to negate the initial positive roll. Hence, in the design of roll damper and yaw damper, a successful safe turn with an acceptable performance must be maintained.

The system ensures a good flying quality over the entire flight regime. Hence, an aircraft designer will achieve an improvement in lateral stability without compromising the level of lateral flying qualities. The actuator for this damping is the aileron (Figure 5.10), and the measurement device is the roll rate gyro to measure the roll rate ($p = \dot{\Phi}$). Any disturbance in roll angle (ϕ) will be damped, so the bank angle will go to zero.

Many air vehicles are provided with a roll rate damper to artificially provide adequate damping in spiral. An aircraft is required to perform a precise mission (e.g., taking g picture or aerial refueling), the roll damper will facilitate the mission. Moreover, the closed-loop control of roll rate is used to reduce the variation of roll performance with altitude/airspeed.

In an aircraft with a conventional configuration, the rolling motion (lateral dynamics) is not decoupled from the yawing motion (directional dynamics). Therefore, the roll augmentation

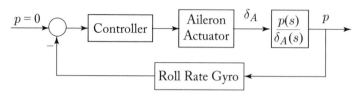

Figure 5.10: Block diagram of a basic roll damper.

system (roll damper) will be simultaneously designed with a yaw augmentation system (yaw damper). Hence, a multivariable state equation (two inputs, ailerons and rudder, and two or more outputs) are employed.

The bank-angle-to-aileron-deflection transfer function has a fourth-order characteristic equation. Using roll approximation, the bank-angle-to-aileron-deflection transfer function is obtained [2] as:

$$\frac{\varphi(s)}{\delta_A(s)} = \frac{L_{\delta_A}}{s^2 - sL_p},$$ (5.20)

where L_{δ_A} is the aileron control derivative and L_p is the lateral stability (roll damping) derivative.

The bending moment generated by the ailerons are transmitted through the wing flexible-beam structure. Since this effect is sensed by the roll-rate gyro in the fuselage, a low-frequency bending mode filter is needed to prevent any response by roll damper due to wing bending. The bending-mode filter is designed to compensate for such changes.

Landing approach takes place at a relatively high angle of attack (about 10–12°). More-over, satisfactory damping of the Dutch roll mode is significant for a safe landing during gusty crosswind. A roll-rate feedback is helpful to ensure good roll response during approach. Thus, a roll damper is very beneficial during landing operation. The controller is designed for a satisfac-tory damping of the Dutch roll mode.

Another challenge in the yaw damper design process is to determine the con-troller/compensator optimum transfer function. Due to significant roll-yaw cross-coupling, there are difficulties in obtaining a good roll response at low dynamic pressure and high an-gle of attack. If the controller gain is too high, it will be excessive at lower angles of attack. However, a high value will cause the aileron servos to reach their maximum rate, and deflection limits rapidly, as they become less effective at the higher angles of attack. Thus, a right value will provide an artificially laterally stable aircraft, while delivering an acceptable roll performance. If an aircraft has an excellent inherent roll damping (e.g., most modern transport aircraft), it will not need a roll damper in inner loop of the bank angle control system.

5.5   LATERAL-DIRECTIONAL SAS

Another more effective solution to a low directionally stable aircraft problem is to augment the directional stability by employing one feedback for each of the x and z axes. Body-axis roll

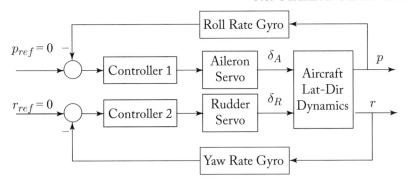

Figure 5.11: Block diagram of a lateral-directional SAS.

rate is fed back to the ailerons, while yaw rate is fed back to the rudder (Figure 5.11). In this yaw damper, the coupling between the rolling and yawing motions is considered in the design process. Indeed, this SAS configuration augments both lateral stability and directional stability simultaneously.

This SAS configuration no only damps the Dutch roll mode, but also modifies the roll-subsidence mode (via roll rate feedback). Moreover, it reduces the variations of roll performance with flight conditions in coordinated turns. This is particularly significant at landing approach, where it takes place at a relatively high angle of attack, and there may be gusty crosswind.

In a basic lateral-directional SAS, there are two feedbacks: (1) roll rate (p), which is measured by a roll rate gyro; and (2) yaw rate (r), which is measured by a yaw rate gyro. Hence, there will be two controllers, one for aileron, and one for rudder.

Figure 5.12 demonstrates the results of a simulation to investigate the response of a fighter aircraft to a doublet pulse as the input. In Figure 5.12a, an aileron doublet ($\pm$1 deg) was applied to the aircraft. Note the difference between the roll rate response of the open-loop dynamics (lateral-directional SAS off) with the closed-loop response (lateral-directional SAS on). In Figure 5.12b, a rudder doublet ($\pm$1 deg) was applied to the aircraft. Note the difference between the yaw rate response of the open-loop dynamics (lateral-directional SAS off) with the closed-loop response (lateral-directional SAS on).

The figure illustrates a major improvement in Dutch roll damping, and a considerable improvement in yaw rate response.

5.6   STALL AVOIDANCE SYSTEM

The stall angle (α_s) is the angle of attack (AoA) at which the airfoil stalls, i.e., the lift coefficient will no longer increase with increasing angle of attack. The stall angle is directly related to the flight safety, since the aircraft will lose the balance of forces in a cruising flight. If the stall is not controlled properly; the aircraft may enter a spin and eventually crash. A practical solution of

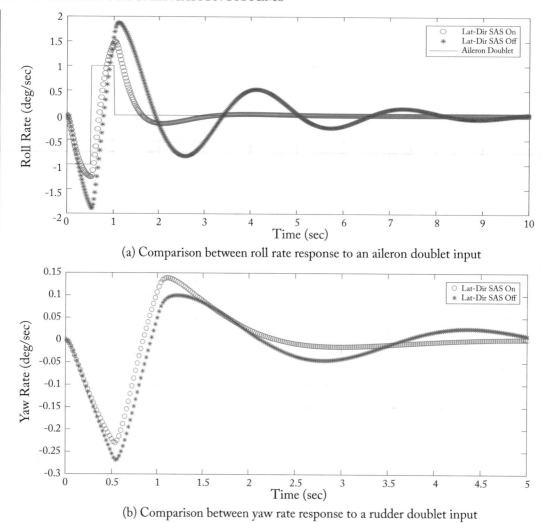

(a) Comparison between roll rate response to an aileron doublet input

(b) Comparison between yaw rate response to a rudder doublet input

Figure 5.12: Yaw rate and roll rate responses to doublet inputs.

the pitch up problem is to limit the aircraft to angles of attack below the stall angle of attack. For an aircraft that is subject to pitch up at high angles of attack, the ability of this system to control pitch up and avoid stall must be checked.

Aircraft stall—recognized as a critical safety aircraft problem—is a situation that exists when the lift of the aircraft wing is less than the apparent weight of the aircraft. As a result, stall warning and avoidance systems are currently in use in most aircraft systems. During a stall,

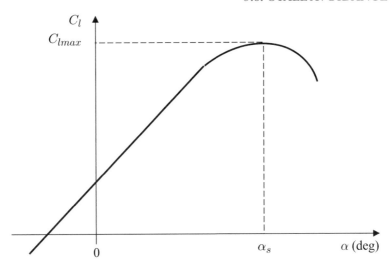

Figure 5.13: The variations of lift coefficient vs. angle of attack.

any one (or a combination) of the following characteristics may be experienced: lack of pitch authority; lack of roll control; buffeting; inability to arrest descent rate; and steep dive.

Figure 5.13 shows the typical variations of lift coefficient vs. angle of attack for a positive cambered airfoil. The typical stall angles for majority of wing airfoils are between 12–16°. This means that the pilot/autopilot is not allowed to increase the wing angle of attack more than about 16°.

The stall may happen for two reasons: (1) the angle of attack exceeds the stall angle of attack; and (2) airspeed is reduced below stall speed. The stall angle of attack is mainly a function of wing configuration (e.g., high lift device deflection) and Reynolds number. However, the stall speed is a function of aircraft weight, aircraft configuration, (e.g., wing and tail) acceleration, engine thrust/power, and atmospheric conditions. An aircraft stall avoidance system uses flight measurements in computing the speed at which a specific aircraft is likely to stall in real time.

Other names for such system are: (1) Automatic Stall-Prevention System; (2) Automatic Protection System; and (3) Maneuvering Characteristics Augmentation System (MCAS). The last one is designed by Boeing as a mode of a flight control system that attempts to mimic pitching behavior, especially in low-speed and high AoA flight. It initiates a nose down (Figure 5.14) above a threshold AoA (say $\alpha_w - 2°$). The system is originally designed to prevent the airplane stall, when making a steep turn under manual control. It adjusts horizontal stabilizer trim to bring the aircraft nose down when it detects that the aircraft is in imminent danger of entering a stall. The system should be always on and to provide continuous stall avoidance information to the pilot along with positive warning signals when critical safety margins are passed.

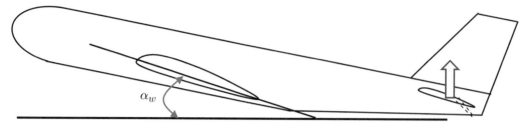

Figure 5.14: Stall control via elevator down-deflection.

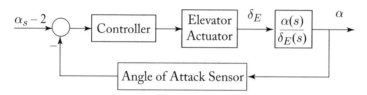

Figure 5.15: Block diagram of an angle of attack control system.

The aircraft stall angle (α_s) is a function of aircraft configuration, type of high lift device, high lift device deflection, and wing configuration. It can be obtained by dividing the maximum lift coefficient ($C_{L_{\max}}$) to the aircraft lift curve slope (C_{L_α}):

$$\alpha_s = \frac{C_{L_{\max}}}{C_{L_\alpha}}. \tag{5.21}$$

Reference [19] has developed and patented a stall avoidance system using flight measurements such as accelerations, aircraft configuration, engine power, and atmospheric conditions. It is interesting to note that the MCAS (anti-stall system) was found by NTSB to be involved in the crash of two Boeing 737 Max in 2019. Boeing proposed changes to the Boeing 737 Max MCAS to address these unfortunate incidences. Since the MCAS design flaws were linked to both crashes, hundreds of B-737 max were grounded until the MCAS and respective sensors are fixed.

The general block diagram of a stall avoidance system is shown in Figure 5.15. The measurement device is an angle of attack sensor, and the control input is applied via the elevator up-deflection that commands nose-down.

The angle of attack sensor is often a small airflow-sensing vane located on the leading edge of an aircraft wing. It is difficult to accurately measure the direction of the local airflow. The problem is to determine whether the direction of the local airflow measured is the direction of the undisturbed air or the distorted local airflow. It is interesting to note that the main reason behind the fatal crash of the Boeing 737 Max in 2019 was a faulty angle of attack sensor.

In more sophisticated, high-performance aircraft, it includes a wing-mounted lift transducer and angle of attack indicator which displays the aircraft angle of attack in terms of degrees or percentages of stall angle, or referenced to a desired angle-of-attack. More details for angle of attack control system are provided in Section 5.2, they are not repeated here.

5.7 AUTOMATIC CONTROL OF TRIM WHEEL

Trim is one of the inevitable requirements of a safe flight. When an aircraft is at trim, the aircraft will not rotate about its center of gravity (cg), and the aircraft will keep moving in a desired direction (e.g., cruise). The horizontal tail is responsible to maintain longitudinal trim and make the summation of moments to be zero, by generating a necessary horizontal tail lift and contributing in the summation of moments about y-axis. An aircraft needs to be re-trimmed after any pitch/power/configuration change. Thus, with automatic trim, the flight is more enjoyable and less fatiguee to the pilot.

The automatic control of trim wheel (sometimes referred to as the speed trim system) is presented in this section. The speed trim system is a type of speed stability augmentation system designed to improve flight characteristics during flight operations with a low aircraft weight, aft center of gravity, and high thrust. The automatic trim is frequently part of automatic Fly By Wire (FBW) system in modern large transport aircraft. It provides longitudinal trim input (i.e., elevator tab deflection) to the horizontal tail. Since, this is a new feature for modern autopilots, more details are provided.

The application of the longitudinal trim equation leads to the following:

$$\sum M_{cg} = 0 \Rightarrow M_{owf} + M_{L_{wf}} + M_{L_h} = 0. \tag{5.22}$$

To expand the equation, we need to define (see Figure 5.16) the variables of wing-fuselage lift (L_{wf}), horizontal tail lift (L_h), and wing-fuselage aerodynamic pitching moment (M_{owf}). We can substitute the values of wing-fuselage and horizontal tail moments into the Equation (5.22):

$$M_{owf} + L_{wf} \left(h\overline{C} - h_o\overline{C} \right) - L_h \cdot l_h = 0. \tag{5.23}$$

The non-dimensionalized form of the longitudinal trim equation is obtained as:

$$C_{m_{owf}} + C_L (h - h_o) - \frac{l}{\overline{C}} \frac{S_h}{S} C_{L_h} = 0. \tag{5.24}$$

At any airspeed, aircraft weight, aircraft center of gravity, altitude, and configuration, the horizontal tail lift coefficient (C_{Lh}) is responsible to make this equation equal to zero. The horizontal tail lift coefficient is changed by pilot/autopilot via elevator deflection. In general, control surfaces (e.g., elevator, aileron, and rudder) are controlled by a human pilot or an autopilot. The link to the control surfaces is commonly by use of the elevator/aileron, stick/yoke/wheel, and rudder pedals. The elevator deflection will generate a hinge moment; this moment is quite large in large transport aircraft.

The hinge moment created by a control surface must be such that the pilot is capable of handling the moment comfortably, as well as the effort should also be small enough to ensure that

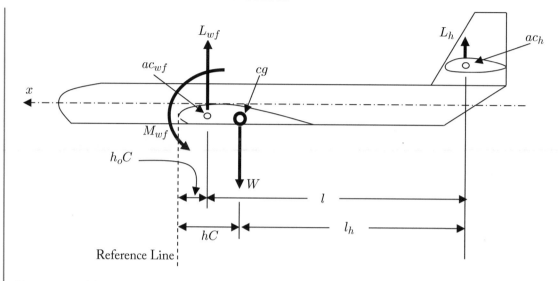

Figure 5.16: The distance between cg, ac_h, and ac_{wf} to the reference line.

the pilot does not tire in a prolong application (say a few hours). Furthermore, the aerodynamic force produced by a control surface; if not managed properly; may interact with inertia and generate an undesirable structural phenomenon called flutter. The control of elevator and tail incidence is performed in the cockpit via rotating the trim wheel.

There are two methods to reduce/nullify the yoke force (i.e., aerodynamic balance): (1) varying the tail angle of attack (i.e., incidence); and (2) varying the trim (usually servo) tab. In general, the second method will lead to a lower trim drag, while requires a more complex mechanism. Aerodynamic balance of control surfaces is frequently addressed for all control surfaces such as elevator, aileron, and rudder. In large transport aircraft (e.g., older Boeing aircraft), an adjustable horizontal tail is generally used for the purpose of setting yoke force equal to zero. In fact, modern Boeing 737s do not have trim tabs, they are equipped with servo tabs for when the flight controls are in "Mechanical mode." For both cases, the cockpit elevator trim wheel is provided.

The yoke control force and respective hinge moment is usually very large in airliners. In some multiengine transport airplanes, it taxes the pilot's strength to overpower an improperly set elevator trim tab on takeoff or go-around. He/she then re-trims the elevator to neutralize the force and continues with the flight.

The hinge moment created by a control surface is defined similar to other aircraft aerodynamic moments as:

$$H = \frac{1}{2}\rho U_1^2 S_c C_c C_h, \qquad (5.25)$$

where S_c denotes the planform area of the control surface (e.g., S_E), and C_c denotes the mean aerodynamic chord of the control surface (e.g., C_E). The parameter C_h is the hinge moment coefficient and is given by:

$$C_h = C_{h_o} + C_{h_\alpha} \alpha_{LS} + C_{h_{\delta_c}} \delta_c, \tag{5.26}$$

where α_{LS} is the angle of attack of the lifting surface (e.g., tail), δ_C is the control surface deflection (e.g., δ_E), and δ_t is the tab deflection. Control surfaces may be aerodynamically balanced without a nose treatment; by employing a tab at the trailing edged of the control surface. When a lifting surface is equipped with a tab, the hinge moment coefficient; C_h is given by:

$$C_h = C_{h_o} + C_{h_\alpha} \alpha_{LS} + C_{h_{\delta_c}} \delta_c + C_{h_{\delta_t}} \delta_t. \tag{5.27}$$

There are a number of tabs used in various aircraft in order to considerably reduce the hinge moment and control force. The most basic tab is a trim tab; as the name implies, it is used on elevators to longitudinally trim the aircraft in a cruising flight. Trim tabs are used to reduce the force the pilot applies to the stick to zero. Tab ensures that the pilot will not tire for holding the stick/yoke/wheel in a prolonged flight. Trailing edge tabs are employed as variable trimming devices, operated by trim wheel directly from the cockpit.

The tab deflection is proportional to the control surface deflection. In general, trim tab serve two functions: (1) it provides the ability to zero-out the stick/wheel/yoke force; and (2) it provides aircraft speed stability at the trim speed. Large transport aircraft such as Boeing 747 and Boeing 737 are equipped with horizontal tail trim tab, as well as the vertical tail trim tab. Military transport aircraft Lockheed C-130B is equipped with horizontal tail trim tab, as well as the servo tab.

The deflection of the control surface can also be performed via servo tab. In another word, pilot controls the servo tab, but servo tab controls the control surface. The stick/wheel force depends on the hinge moment of both control surface and the tab. A servo tab is a tab in which the stick/wheel is connected directly to the tab, which is hinged to the control surface. Similar to other tabs, the main function of the servo tab is to reduce the pilot force when the stick/wheel is moved.

Figure 5.17 illustrates two methods for controlling horizontal tail and elevator via trim wheel: 1. Fixed-tail, elevator, and trim tab, 2. Adjustable tail and elevator. In both cases, the trim wheel can be rotated by human pilot, or be motorized. An electric motor can be controlled by autopilot to adjust its position to locate either tail or trim tab. In either case, the aircraft must be equipped with trim cutout switches, one for the autopilot, and one for the normal electric trim. Using these would immediately disable the electric trim. These should be generally part of a standard trim runaway situation.

The elevator is for direct, fast, and limited authority control of the aircraft pitch attitude and angle of attack. However, the tail is for trimming the aircraft position; when the yoke/stick is release, it will return to the zero force position. It will eventually leads to a steady state flight.

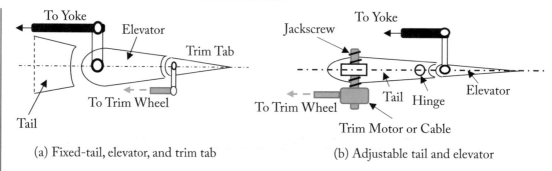

(a) Fixed-tail, elevator, and trim tab

(b) Adjustable tail and elevator

Figure 5.17: Controlling horizontal tail via trim wheel.

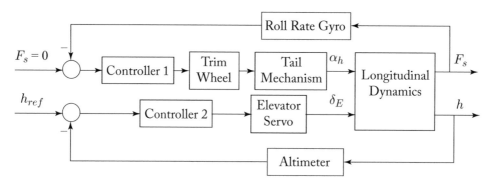

Figure 5.18: Block diagram of automatic control of trim wheel.

In cases a, when the cockpit trim wheel is moved to nose down airplane, the trailing edge of the elevator trim tab is moved up. With the same token, when the cockpit trim wheel is moved to nose up airplane, the trailing edge of the elevator trim tab is moved down. The elevator trim could be actuated either manually/automatically by movement of the trim wheel located to the sides of the throttle pedestal, or electrically by an electric trim button located on the left side of the pilot's control yoke. The trim wheel is adjusted/rotated to unload the elevator.

In case b, the trim wheel is directly controlling the horizontal tail setting angle. The horizontal tail has a rear fulcrum/pivot/hinge and front jackscrew. The yoke deflects the elevators, whereas the trim wheel deflects the horizontal tail. The elevator, positioned by yoke, is part of the tail. If the yoke is pulled back, the elevator moves up and fuselage nose will pitch up. Then, the tail must deflect downward, either automatically or manually, to remove the yoke loading. The tail is much larger than the elevators, and therefore is more effective than the elevator at extreme angles.

Figure 5.18 demonstrates the block diagram of a speed stability augmentation system, or automatic control of trim wheel, for the case, where the trim wheel controls the tail setting angle (α_h). The net result of control system is that the aircraft maintains both altitude and zero yoke force. There are two feedbacks: (1) yoke force (F_s), which is measured by a force sensor; and (2) altitude, which is measured by an altimeter. Hence, there will be two controllers, one for elevator, and one for trim wheel. The main commanded reference is a zero yoke force, and the actuator is a motorized trim wheel. For both feedback signals, filters (not shown in the figure) may be employed to filter the noise.

The yoke force is a function of hinge moment and the power transmission mechanism:

$$F_s = G_e H. \tag{5.28}$$

The parameter G_e is the gearing ratio and has a unit of rad/in or deg/m. The acceptable range for the yoke force (for both temporary and prolonged flight) is governed by FAA regulations. As the size of the aircraft (in fact, horizontal tail planform area), airspeed, and tail setting angle are increased, the hinge moment and yoke force are increased too. Noting that the tail angle of attack is a function of downwash (ε), and plugging hinge moment from Equation (5.25) into Equation (5.28), yields [2]:

$$F_s = \frac{1}{2}\rho U^2 S_e C_e G_e \left[C_{ho} + C_{h\alpha} \left\{ \alpha \left(1 - \frac{d\varepsilon}{d\alpha} \right) + \alpha_h - \varepsilon_o \right\} + C_{h\delta E}\delta_E \right], \tag{5.29}$$

where S_e, C_e, are elevator planform area and elevator chord, respectively. A compilation of wind tunnel investigation data for elevator hinge moment coefficient (C_h) is presented by [18] for models of nine related horizontal tails. These experimental results can be used to determine the accuracy of estimating the hinge moment. Using Equation (5.28), one can derive the yoke-force-to-tail-setting-angle transfer function (i.e., $F_s(s)/\alpha_h(s)$).

5.8   QUESTIONS

5.1. Name three aircraft, which are equipped with at least one form of stability augmentation system.

5.2. Draw a general simplified block diagram of a stability augmentation system.

5.3. Name at least four functions of stability augmentation system.

5.4. Briefly describe the basics of a roll damper.

5.5. Briefly describe the basics of a pitch damper.

5.6. Briefly describe the basics of a yaw damper.

5.7. Briefly describe the basics of an automatic control of trim wheel.

5.8. What is the primary function of a pitch damper?

5.9. What is the primary function of a roll damper?

5.10. What is the primary function of a yaw damper?

5.11. What stability derivatives are improved by a roll damper?

5.12. What stability derivatives are improved by a pitch damper?

5.13. What stability derivatives are improved by a yaw damper?

5.14. Draw the block diagram of a basic roll damper.

5.15. Draw the block diagram of a basic pitch damper.

5.16. Draw the block diagram of a basic yaw damper.

5.17. Draw the block diagram of an angle of attack control system.

5.18. Draw the block diagram of an automatic control of trim wheel.

5.19. Briefly describe the features of an angle of attack sensor.

5.20. Briefly compare a pitch damper and a pitch-axis SAS.

5.21. Briefly compare a yaw damper and a lateral-directional SAS.

5.22. What sensor is employed to measure the pitch rate?

5.23. Name five Navy tactical aircraft that were equipped with SASs in 1980s.

5.24. What oscillatory motion modes are targeted by a pitch damper for improvement?

5.25. What oscillatory motion mode is targeted by a yaw damper for improvement?

5.26. Characterize a first-order high-pass washout filter.

5.27. Suggest a solution to address the inherent static directional stability problems of tailless fighter aircraft.

5.28. When an aircraft does stall?

5.29. What are typical stall angles for majority of wing airfoils?

5.30. What are other names for the stall avoidance system?

5.31. Describe the reasons behind fatal crashes of two Boeing 737 Max in 2019.

5.32. What is the function of a trim wheel?

5.33. What are two methods to reduce/nullify the yoke force in a transport aircraft?

5.34. What parameters do influence the hinge moment of an elevator?

5.35. Describe the technique to control the horizontal tail via a trim wheel.

CHAPTER 6

Command Augmentation Systems

6.1 INTRODUCTION

Some autopilot modes improve the stability of the aircraft (i.e., SAS, as covered in Chapter 5), while some modes enhance the aircraft's maneuverability. These modes augment the aircraft response to a control input. The slow motion modes (e.g., phugoid in longitudinal motion, and spiral in lateral-directional motion) can be controllable by a human pilot. Since it is undesirable for a pilot to pay continuous attention to controlling these modes, some autopilot modes are dedicated to help pilots to create an enjoyable flight. The command augmentation system (CAS) is intended to provide the pilot with a particular type of response to the control inputs. Hence, the Command Augmentation Systems are autopilot modes and are a type of automatic flight control system to provide "pilot relief;" particularly in high performance fighters. Note that the command signal is primarily coming from the pilot.

In another term, the CAS is intended to control the aircraft natural motion modes, and to provide the pilot with a particular type of response to the control inputs. In [1], the SAS and CAS are referred to as non-autopilot functions, while hold functions are referred to as the autopilot modes. In this, text, the SAS and CAS, hold modes, and flight path control modes are all referred to as the AFCS functions. In many cases, both SAS and CAS control an angular rate. However, the SAS controls the angular rates when created by an atmospheric disturbance, the CAS controls the angular rates when created by a pilot/autopilot.

Reference [3] refers to CAS to "Active control technology (ACT)" to use a multivariable AFCS in order Control technologies have progressed to the point where the AFCS can provide the pilot with selectable task tailored functions. The CAS is employed to improve the maneuverability, the dynamic flight characteristics and, the structural dynamic properties of an aircraft. Employing a CAS, an aircraft produces a degree of maneuverability beyond the capability of a conventional aircraft.

Moreover, the structural loads which the aircraft would have experienced as a result of its motion without an CAS system are much reduced. Without CAS, an aircraft may experience the repeated high levels of stress, and the peak loads to which the aircraft is subjected. These structural vibrations can impair the life of the aircraft structure; hence, the CAS will provide the suppression of flutter. Another benefit of CAS is to prevent severe penalties upon the aircraft's flight operation, mostly in terms of reduced payload, or reduced performance.

In the category of command augmentation, three basic modes are available: (1) command tracking; (2) command generator tracker; and (3) normal acceleration CAS. The command tracking system are mainly divided into two modes: (1) pitch rate tracking and (2) roll rate tracking. Command generator tracker is known as the model following system, since it creates time varying trajectories. The CAS provides augmentation to all longitudinal, lateral, and directional control functions.

One of the basic CAS for highly maneuverable aircraft is the relaxed static stability system. In theory, the aircraft's maneuverability can be enhanced by relaxing static (longitudinal/lateral/directional) stability. A better dynamic response to the control surface deflections is also provided by relaxing static stability system. In practice, this AFCS will result in a reduction in the trim drag, while make the tail area smaller. Moreover, a considerable reduction in aircraft structural weight can be achieved. When the aircraft static stability is relaxed, the dynamic stability is downgraded. In designing the Relaxed Static Stability System (RSSS), the designer must make sure that the aircraft's handling qualities are restored.

An AFCS mode that stabilizes the aircraft while providing automatic turn coordination and pitch-roll command tracking is usually referred to as a CAS. Often a linear control law is employed, and gains are scheduled with respect to key flight variables such as Mach number, load factor, and altitude. Sensor/actuator nonlinearities such as saturation, dead-zone and hysterisis should also be accommodated in this framework. To provide robustness, compensators should be employed that provide desired disturbance rejection properties. Reference [42] presents the development of a nonlinear command augmentation system for a high performance fighter.

In this chapter, objectives, block diagrams, and features of four command augmentation systems are presented: (1) normal acceleration CAS; (2) pitch rate CAS; (3) lateral-directional CAS; and (4) Gust Load Alleviation System (GLAS).

6.2   NORMAL ACCELERATION CAS

In order to enhance the fighter aircraft utility in combat maneuvers, a normal acceleration (via roll rate, p) command augmentation system is often employed. The normal acceleration (a_n) is the component of acceleration in the negative direction of the body-fixed z-axis. At a small angle of attack (α), the lift direction is nearly coincident with the body negative z-axis. In aircraft applications, the acceleration is often measured in unit of g. As discussed in Chapter 3, the ratio of the lift (L) by the aircraft weight (W) is called the load factor (n):

$$n = \frac{L}{W}. \tag{6.1}$$

Other than turning maneuver, the normal acceleration is generated during pull up and push over maneuvers. In low α, β, and ϕ the accelerometer output is an approximate measurement of load factor. The normal acceleration (i.e., accelerometer output) is determined, when 1 g is subtracted from the load factor:

$$a_n \approx n - 1. \tag{6.2}$$

Consider an aircraft is pitching, but neither yawing nor rolling. The normal acceleration is not usually measured at the aircraft's c.g., but at some other stations. Consider an accelerometer installed on the body x-axis, at a distance l_a forward of the aircraft center of gravity (cg). The normal acceleration—measured by such accelerometer—is a function of linear acceleration in z-direction ($\dot{w} = a_z$) and pitch angular acceleration about the cg:

$$a_n = \dot{Q} l_a - a_z. \tag{6.3}$$

In the case of a constant airspeed (U_o) and a pitching motion, the normal acceleration is obtained by:

$$a_n = U_o (Q - \dot{\alpha}) - a_z, \tag{6.4}$$

where Q is the pitch rate. In general, the normal acceleration is a function of airspeed, angle of attack, pitch rate, and elevator deflection. Now that, the normal acceleration and load factor are defined, a few applications of the normal acceleration CAS is briefly described.

In any turning flight, a normal acceleration is generated due to the bank angle. In a maximum rate turn and minimum turn radius turn, the g-load must not exceed the structural limit (i.e., load factor) of the aircraft, or the pilot's physical limits. A normal acceleration CAS is intended to help the pilot to control the acceleration generated along the negative z-axis.

The maximum allowable load factor is limited by the aircraft structural strength, or human tolerance. Every aircraft has a unique maximum allowable load factor. For instance, the Eurofighter Typhoon (Figure 6.1)—a twin-engine, canard-delta wing, multirole fighter—has a maximum load factor of 9. Human pilot and passengers have a limit handling g-load, since the blood circulation is affected by number of g's (i.e., load factor). In a pull-up maneuver, the pilot's concern is with the steady state normal acceleration. A regular human/pilot feels uncomfortable when the load factor is exceeding 2. The turn would stress a pilot if it involves a sustained normal acceleration of 2 g's or higher.

Another application of a normal acceleration CAS is to protect flight attendants from discomfort in a long fuselage. Wind gusts tend to have a considerable impact on the normal acceleration. A long slender fuselage may experience lateral and vertical accelerations in the forebody having a natural frequency of about 1 Hz. This vibration frequency will cause a notable discomfort to pilot and passengers. In this situation, a suitable controlled variable for the longitudinal motion is the normal acceleration (n_z) of the aircraft.

Moreover, for the pilot of a high performance fighter—which has to maneuver (e.g., dogfight) the aircraft to its performance limits and to perform challenging tasks such as precision tracking of a target—a normal acceleration CAS provides more efficient mission. Hence, for high-performance turns, a normal acceleration (i.e., g-command) is an appropriate reference command of operation of the flight control system. In the above cases, it is required in AFCS work to control the normal acceleration at the pilot's station, or height.

Figure 6.2 illustrates the block diagram of a normal acceleration CAS. There are three feedbacks (normal acceleration, pitch rate, and rate of pitch rate), and two sensors (normal ac-

Figure 6.1: Royal Air Force Eurofighter EF-2000 Typhoon F2 (courtesy of Tim Laurence).

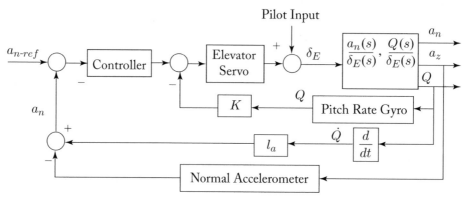

Figure 6.2: Block diagram of a normal acceleration CAS.

celerometer and pitch rate gyro). The parameter "l_a" in the second loop represents the moment arm for the accelerometer. If the accelerometer is located at the vehicle's cg, the l_a will be zero, and the a_n and n_z will be the same. The system has two inputs: (1) reference a_n and (2) pilot input via stick. Thus, the elevator is deflected by two sources in parallel. One is the pilot for intended maneuver, and one is the control system to compensate the shortcoming of the aircraft response in normal acceleration.

The normal-acceleration-to-elevator-input transfer function has a fourth-order characteristic equation:

$$\frac{a_n(s)}{\delta_E(s)} = \frac{K\left(s^3 + b_2 s^2 + b_1 s + b_0\right)}{a_4 s^4 + a_3 s^3 + a_2 s^2 + a_1 s + a_o}. \tag{6.5}$$

The normal acceleration is not a state variable, the expressions for the numerator and denominator coefficients can be derived by expanding the longitudinal state-space model. Employing the angle of attack (α), the normal-acceleration-to-elevator-deflection transfer function may be obtained by:

$$\frac{a_n(s)}{\delta_E(s)} = \frac{a_n(s)}{\alpha(s)} \frac{\alpha(s)}{\delta_E(s)}. \tag{6.6}$$

The angle-of-attack-to-elevator-deflection transfer function ($\frac{\alpha(s)}{\delta_E(s)}$) has been introduced in Chapter 3. The normal-acceleration-to-angle-of-attack transfer function [3] is a function of a number of longitudinal stability and control derivatives as:

$$\frac{a_n(s)}{\alpha(s)} = \frac{U_o}{g M_{\delta_E}} \left(Z_{\delta_E} M_w - M_{\delta_E} Z_w\right). \tag{6.7}$$

The load factor (n) can be expressed in terms of the rate of change of flight path angle (γ) as:

$$n = \frac{U_o \dot{\gamma}}{g}. \tag{6.8}$$

Since flight path angle is: $\gamma = \theta - \alpha$, the load-factor-to-elevator-input transfer function is obtained as:

$$\frac{n(s)}{\delta_E(s)} = \frac{U_o}{g} s \left(\frac{\theta(s)}{\delta_E(s)} - \frac{\alpha(s)}{\delta_E(s)}\right). \tag{6.9}$$

The normal acceleration at the aircraft's c.g. should be minimized, so that the output response should lie between specific limits. By such initiative, the ride discomfort index will be minimized.

The design of a normal acceleration control system to achieve good handling qualities is a challenging task. Various controllers may be employed to satisfy the normal acceleration control requirements; the simplest ones are just two gains. When all transfer functions are derived, the design problem statement is to determine gain K and controller features. An optimal controller can be designed such that it provides a unique and stabilizing control law. The controllers' gains can be refinement through flight tests.

A disadvantage of normal acceleration feedback is that the gain (K) of the transfer function varies widely with airspeed. An accelerometer is very sensitive to structural vibration. Hence, for accurate flight path control in a turbulent air, a relatively low threshold is required for the accelerometer. Moreover, a certain amount of filtering should be added to the measured signal.

A command augmentation system that yields a good normal acceleration step response may have a pitch rate response with a very large overshoot. Hence, the normal acceleration command augmentation system and the pitch rate command augmentation system should be designed simultaneously.

The control system of fighter aircraft is often designed with the objective to provide **the pilot, a control over normal acceleration at high speed, while pitch rate at low speed**. This outcome tend to provide augmentation over the flight parameter that requires more pilot attention.

6.3 PITCH RATE CAS

Another common mode of operation for a pitch axis command augmentation system is as a pitch rate (Q) command system. There are a number of flight operations and maneuvers, which require a well-behaved pitch rate response. For instance, this system can be employed to stop a pitch-up which often happens so rapid that a pilot is unable to control.

An example of an effective pitch rate CAS is a maneuver which requires a high angle of attack. At a large angle of attack, a strong nose-up pitching moment may be produced. In such flight condition, it is possible that horizontal tail is stalled, hence, it is difficult to generate the necessary opposing pitching moment by elevator deflection. The solution is a command augmentation system to generate a nose-down pitching moment.

There are two conflicting requirements for pitch rate response in a fighter aircraft in two flight operations: (1) high overshoot for a good acquisition of target and (2) no overshoot for a precise target tracking by using of a sighting device. A well-designed pitch axis CAS cab deliver an acceptable pitch rate response in both cases. Control of pitch rate is also the preferred objective during approach and landing.

A pitch rate command augmentation system that yields a good overshoot, may lead to a sluggish normal acceleration response. It is recommended to smoothly blend together the command augmentation of pitch rate and normal acceleration. Hence, the simultaneous design of normal acceleration command augmentation system and the pitch rate command augmentation system will be more beneficial.

In the flight maneuver in the x-z plane, the main control surface is the elevator, and six flight outputs are angle of attack, pitch angle, pitch rate, airspeed, altitude, and normal acceleration. However, variables of pitch rate (q) and normal acceleration (a_n) are two sources of most concern for a fighter pilot. These two variables may be linearly combined [3] to generate a new parameter called C-star (C^*):

$$C^* = x_{pilot} n_z + K_1 Q, \tag{6.10}$$

where Q is in rad/sec, and n_z is in g's, and x_{pilot} is an indication of the pilot's station (the accelerometer distance forward of the cg). The coefficient K_1 is selected based on the aircraft configuration and the satisfactory maneuvering requirements. The response of C^* to an elevator input or gust should have [1] the following features: (1) maximum overshoot, less than 60%; (2) rise time, less than 2 seconds; (3) settling time, less than 3 sec; and (4) zero steady-state error.

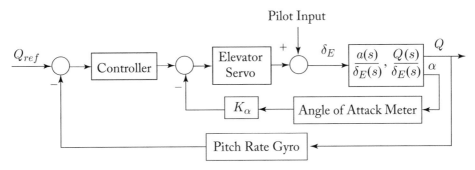

Figure 6.3: Block diagram of a pitch rate CAS.

The C^* can be treated as an output of the aircraft, since it is a function of time. The parameter C^* is an indication for contributions of the normal acceleration, and the pitch rate to the total response of the aircraft. The C^* is zero initially, but builds up to its corresponding steady state value. The transient response of the C-star is characterized by the second-order mode associated with the short period motion.

Figure 6.3 demonstrates the block diagram of a pitch rate CAS that includes an inner loop. The system has two inputs: reference q and pilot input via stick. Thus, the elevator is deflected by two sources in parallel. One is the pilot for intended maneuver, and one is the control system to compensate the shortcoming of the aircraft response in pitch rate. The pitch rate is measured a by a rate gyro, and angle of attack is sensed by an angle of attack meter. The pitch-rate-to-elevator-deflection and angle-of-attack-to-elevator-deflection transfer functions ($\frac{\alpha(s)}{\delta_E(s)}, \frac{Q(s)}{\delta_E(s)}$) have been presented in Chapter 3.

This control system would hold a steady climb (in a pull-up) or dive (in a push-over) with no control stick deflection. This control system will augment aircraft longitudinal dynamic stability by adjusting four poles of short-period and phugoid modes to the desired locations.

The inner-loop feedback controller could be just a gain. The main controller of the lead-lag type (i.e., $\frac{K(s+a)}{s+b}$) can be employed to control both angle of attack and pitch rate. However, a PI controller will provide more precise control. The selection and the design of control law is finalized by using a multidisciplinary optimization technique.

The design can be initially performed on the dynamics of short-period mode. The inner loop angle of attack feedback—with the controller of K_α—is added to improve the pitch rate response. Since there are two outputs (Q and α) and one input (δ_E), using state-space model is more convenient. In matlab, you may use "tf2ss" command to convert two transfer functions to one state space model. A promising design will be gained by a trade-off technique to have a desirable pitch rate closed-loop step-response.

6.4 LATERAL DIRECTIONAL CAS

In high-performance aircraft (e.g., fighters) that should be able to maneuver rapidly at high angles of attack, a lateral-directional command augmentation system is highly recommended. This type of CAS will help the pilot to conduct a fast maneuver, which require rolling and yawing motions. Another application of a lateral-directional CAS is to control a potentially catastrophic phenomenon of the inertia coupling in fighter aircraft. Both cases are presented below.

The first example that indicates the effective role of a lateral-directional CAS is a rapid 90° roll at high angle of attack for a turn. In such a maneuver, since the high angle of attack will immediately be converted to a large sideslip angle, the roll will become counterproductive. This is referred to as adverse sideslip, because it will tend to oppose the initial roll, and this roll will be an inefficient turn entry. Moreover, a large sideslip angle is undesirable, since it reduces the effectiveness of the control surfaces (i.e., aileron and rudder). A large sideslip angle will also generate a large side force; even it may break the vertical tail.

To maintain the initial angle of attack to roll in a turn at high angle of attack, the lateral-directional CAS is activated. An aircraft without a lateral-directional CAS has a greatly degraded roll response at high angle of attack. This command augmentation system is designed to command the aircraft to roll around its stability x-axis (velocity vector) rather than the body x-axis.

The inertia coupling is defined as a resonant divergence in pitch/yaw, when roll rate equals the lower of the pitch/yaw natural frequencies. From Euler's equations of motion (Equation (1.14)), the pitching moment, M_{IC}, due to inertia coupling is obtained by:

$$M_{IC} = \dot{Q} I_{yy} = (I_{zz} - I_{xx}) R P. \tag{6.11}$$

In a fighter with a heavy fuselage and a wing with low aspect ratio, the aerodynamic forces and moments generated by the wing and tail are insufficient to stabilize the aircraft in a high angle of attack roll. Therefore, a rapid roll about the stability x-axis, at large angle of attack, can produce a strong nose-up pitching moment. This fighter configuration at such flight conditions will cause problems in maneuvering, and may lead to the loss of aircraft. A human pilot may not be able to counter the pitching moment due to inertia coupling, but a lateral-directional CAS can be designed to nullify such moment.

The main objective of a lateral-directional command augmentation system is to control a turn in a desirable fashion. A desirable yaw in a turn must be such that the ratio between natural frequency of the zero control (ω_c) to the natural frequency of the Dutch roll (ω_{dr}) is greater than one (see Figure 6.4).

Figure 6.5 exhibits the configuration and essential elements of a lateral-directional CAS for a fighter aircraft. In a basic lateral-directional CAS, there are two feedbacks: roll rate (P), which is measured by a roll rate gyro, and yaw rate (R), which is measured by a yaw rate gyro. Hence, there will be two controllers, one for aileron, and one for rudder.

The aileron is controlled by controller 1 based on the error signal (e_P) from commanded roll rate (p_{com}). The aileron input has also an interconnection to the rudder servo via an aileron-

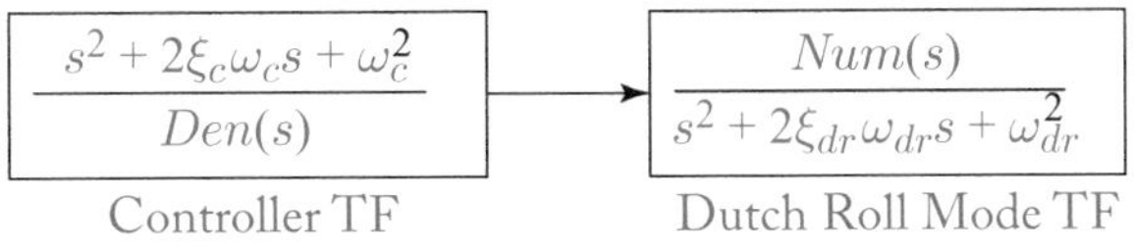

$$\frac{s^2 + 2\xi_c\omega_c s + \omega_c^2}{Den(s)}$$

Controller TF

$$\frac{Num(s)}{s^2 + 2\xi_{dr}\omega_{dr} s + \omega_{dr}^2}$$

Dutch Roll Mode TF

Figure 6.4: Block diagram of a cascade controller to damp Dutch roll.

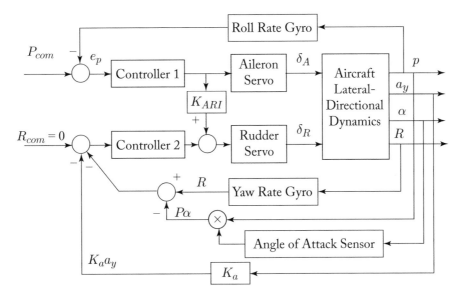

Figure 6.5: Block diagram of a basic lateral-directional CAS.

rudder interconnect (ARI) gain. This gain (K_{ARI}) will be scheduled by the angle of attack (α) and Mach number (M). This interconnection provides the component of yaw rate necessary to achieve a stability axis roll:

$$R = P \cdot \tan(\alpha). \tag{6.12}$$

Thus, in a positive angle of attack, both roll rate (p) and yaw rate (r) must have the same sign. The value of K_{ARI} should be such that to satisfy the yaw rate constraint (Equation (6.12)).

In order to construct the correct feedbacks, we need to convert body-axes roll and yaw rates (P and R) to stability-axes roll and yaw rates (P_s and R_s). Using the rules for finding rotation matrices, the rotation matrices from body to stability, and stability to wind axes are obtained. The stability-axes roll and yaw rates in terms of the body-axes roll and yaw rates (assuming no sideslip) are:

$$P_s = P\cos(\alpha) + R\sin(\alpha) \tag{6.13}$$

$$R_s = -P\sin(\alpha) + R\cos(\alpha). \tag{6.14}$$

Linearizing these equations yields:

$$P_s = P + R\alpha \tag{6.15}$$

$$R_s = -P\alpha + R. \tag{6.16}$$

The command is to have a desired roll rate about stability x-axis, while yaw rate about stability z-axis is zero ($R_{com} = 0$). Moreover, in order to have a coordinated turn, the lateral acceleration (a_y) is required to be zero. Based on these justifications, the error signal (e_R) to drive the controller of the rudder (controller 2) will be:

$$e_R = R_{com} - (R - P\alpha) - K_a a_y. \tag{6.17}$$

In designing controllers, four outputs of roll rate, yaw rate, lateral acceleration, and angle of attack are employed. Three feedbacks are utilized to satisfy Equations (6.17) and (6.13), with acceptable roll rate performance (P_{com}). A roll rate is fed back to aileron path to control the roll rate response. Two feedbacks are employed to control the yaw rate response. The lateral acceleration (a_y) is fed back to yaw rate to cancel the sideswiping. Moreover, the inner feedback loop (yaw rate) in the rudder path provides Dutch roll damping.

The control system should be designed to limit the commanded roll rate (P_{com}) to avoid a pitch up and to limit the side-slipping capability. A fast roll rate response is desired, while spiral mode is stable (with an acceptable time constant), and the Dutch roll is highly damped with high natural frequency. Optimum controllers will cancel out a few poles and zeros to decouple two rudder and aileron channels.

6.5   GUST LOAD ALLEVIATION SYSTEM

6.5.1   ATMOSPHERIC GUST

A considerable atmospheric phenomenon for a flexible aircraft is the disturbance. One of the most important and famous type of disturbances [24] that an aircraft is experiencing is gust. An aircraft may encounter: discrete and continuous atmospheric disturbances/gusts. Despite poor and gusty weather conditions, an aircraft must maintain a specified heading and altitude in order to reach its destination safely. A gust field will influence both the aerodynamic and aeroelastic behavior of an air vehicle.

If vertical gusts are not alleviated, they add stresses to the structure and cause discomfort to the passengers. Control systems are designed for the alleviation of gust loads on a wing using regular symmetrically activated ailerons and similar control surfaces. The primary goal is to alleviate the fatigue and static loads on the wing by reduction of the wing bending moment. The Gust Load Alleviation System (GLAS) has been employed in many aircraft from bomber B-2 Spirit to transport aircraft Airbus 380. Reference [32] has investigated the effectiveness of different control surfaces in the GLAS of a large transport aircraft.

Aeroservoelasticity is a multi-disciplinary area that deals with the interaction of the structural, aerodynamics, and control systems of an aircraft. The interaction impacts various aircraft

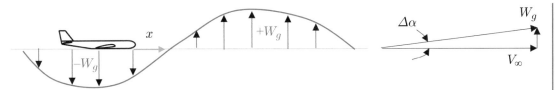

Figure 6.6: Sinusoidal vertical gust.

features from aircraft performance, static and dynamic stability, and controllability, to structural integrity, handling qualities, and airworthiness.

In a bigger picture, flutter, all static and dynamic loads, and their associated stresses and strains should be considered. One of the required tools in GLAS design process is the aeroelastic mathematical model (e.g., beam-type model in MSC/NASTRAN). For instance, in the modal approach of aeroservoelasticity analysis, it is assumed that the structural displacements in response to external excitation (e.g., gust) can be expressed as a linear combination of a relatively small set of displacement modes. Any approximation approach will reduce the accuracy in subsequent analysis, relative to methods that do not require approximations. To read more on the aeroservoelastic gust response analysis for the design of transport aircraft, you may refer to [25].

6.5.2   VERTICAL GUST

A vertical gust impacts the aircraft's normal acceleration. A typical discrete gust has often a period of about less than a second. The purpose of a Gust Load Alleviation (GLA) system is to reducing the effect of gust loads on the aircraft structure (e.g., reducing the wing bending moments) by deflecting control surfaces. The GLA is a technique to control the contribution of the rigid body and the wing bending moments to the aircraft behavior when a gust is encountered. By alleviating gust loads, the ride comfort for the crew or passengers will be improved through reducing the transient peak loads. This—in turn—reduces the structural fatigue.

Figure 6.6 demonstrates a UAV experiencing a vertical sinusoidal gust (i.e., in the z direction). The variation of the vertical gust velocity along the x-axis of the airplane is similar to the velocity distribution created on a pitching airplane. A discrete (or individual) gust (W_g or W_W) idealization consists [31] of a "one-minus-cosine" profile:

$$W_g = \frac{1}{2}U_{de}\left(1 - \cos\left(\frac{2\pi x}{2H}\right)\right), \tag{6.18}$$

where H is the scale and denotes gradient distance ($12.5 \times \overline{C}$ (times wing chord)), and U_{de} represents the derived equivalent gust velocity in fps. The parameter x is the forward aircraft displacement, and may be replaced with "V.t". Hence, the gust model is obtained as:

$$W_g = \frac{1}{2} U_{de} \left(1 - \cos \left(\frac{2\pi V}{2H} t \right) \right).$$ (6.19)

According to FAR 23 [29] and FAR 25 [30], the values of the derived equivalent gust velocity (U_{de}) are specified as functions of aircraft speed and altitude.

The discrete gust is usually considered either as an up-gust, or as a side-gust. In other words, the direction of the gust is perpendicular to the direction of flight. This gust changes the airspeed (in the y or z direction) of the aircraft suddenly. Reference [30] recommends values for the gust velocities at various altitudes. When a stability augmentation system is included in the analysis, the effect of any significant system nonlinearities should be accounted for, when deriving limit loads from limit gust conditions. If the gust field wavelength is large in comparison with the wing span, the gust produces a spanwise variation of velocity along the vehicle.

6.5.3   GUST LOAD MATHEMATICAL MODEL

Atmospheric gust may exist in all three axes (x, y, and z), as sudden airspeed of U_w, V_w, and W_w. The gust along y-axis, side gust (V_w) may hit the nose or vertical tail. In both cases, they are usually handled by the yaw damper via damping any induced sideslip angle and induced yaw rate. A gust along x-axis (U_w) will induce an angle of attack and will generate a pitching moment. This gust output is handled through a pitch damper by damping any change in pitch angle and pitch rate.

The forces associated with gust will be entered [28] the aircraft dynamic model through the moment equations:

$$F_{w_x} = m \left(\dot{U}_w + Q W_w - R V_w \right)$$ (6.20)
$$F_{w_y} = m \left(\dot{V}_w + R U_w - P W_w \right)$$ (6.21)
$$F_{w_z} = m \left(\dot{W}_w + P V_w - Q U_w \right),$$ (6.22)

where the subscript w indicates wind (i.e., gust). These forces are considered as an additional applied forces acting on the aircraft along with weight (F_G), aerodynamics force (F_A), and thrust forces (F_T) as:

$$m \left(\dot{U} + Q W - R V \right) = F_{G_x} + F_{w_x} + F_{A_x} + F_{T_x}$$ (6.23)
$$m \left(\dot{V} + U R - P W \right) = F_{G_y} + F_{w_y} + F_{A_y} + F_{T_y}$$ (6.24)
$$m \left(\dot{W} + P V - Q U \right) = F_{G_z} + F_{w_z} + F_{A_z} + F_{T_z}.$$ (6.25)

6.5.4   WING ROOT BENDING MOMENT

In general, a vertical gust (W_w) may hit two locations, so it might cause two disturbances: (1) if it hits the nose or horizontal tail, it will induce a yawing moment; and (2) if it hits the wing tip or tail tip, it will induce a local angle of attack, $\Delta\alpha$ (see Figure 6.6) and a rolling moment. In the first place, an induced yawing moment is nullified by a yaw damper; and an induced rolling moment is nullified by a roll damper. However, a gust, which hits the wing tip, will increase the wing bending moment at root. The main objective of a gust alleviation system is to handle such wing bending moment. The induced angle of attack (see Figure 6.6) is a function of the gust speed and aircraft airspeed (V_∞):

$$\Delta\alpha = \tan\left(\frac{W_g}{V_\infty}\right). \tag{6.26}$$

Since the gust speed is much less than the airspeed, the induced angle of attack is small. Thus, Equation (6.20) can be linearized:

$$\Delta\alpha = \frac{W_g}{V_\infty}, \tag{6.27}$$

where $\Delta\alpha$ is in radian. Another output of a vertical gust (W_w) is a plunging motion, since there is a rate of change of velocity component in z-axis ($\dot{W}_g$ or simply $\dot{w}$). Hence, there will be a rate of change of attack ($\dot{\alpha}$):

$$\dot{\alpha} = \frac{d\alpha}{dt} = \frac{\dot{w}}{U_1}. \tag{6.28}$$

Both induced angle of attack and rate of change of angle of attack will cause a change in the local lift (ΔL) at the wing tip:

$$\Delta L = qS\Delta C_L, \tag{6.29}$$

where the change in the local lift coefficient (ΔC_L) is:

$$\Delta C_L = C_{L_\alpha}\Delta\alpha + C_{L_{\dot{\alpha}}}\dot{\alpha}. \tag{6.30}$$

Moreover, the change in vertical speed due to a vertical gust ($\dot{W}$) is translated as the local change in the vertical acceleration (a_z). This acceleration can be measured by an accelerometer installed at the wing tip (Figure 6.7).

The bending moment (M_x) at the wing tip will be a function of both wing regular lift (L), and the local wingtip induced lift (ΔL):

$$M_x = \frac{L}{2}y_L + \Delta L y_{\Delta L}, \tag{6.31}$$

where y_L is the distance between wing lift at left or right section to the wing root, and $y_{\Delta L}$ is the distance between local induced lift (due to vertical gust) and the wing root. The 1/2 in Equation (6.31) is because each wing section (left or right) is generating 50% of the total lift. The wing root bending moment is reduced by controlling the normal acceleration at the wing tip. Since the outer ailerons are located at the 80–100% of wing span, they create a large wing bending moment.

The regular lift (L) is a static load for the wing, while the induced local lift (ΔL) is considered as a dynamic load. The wing spar structural failure due to the static load will never happen,

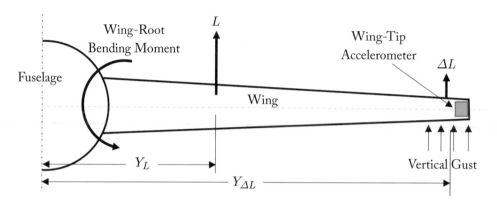

Figure 6.7: Wing-tip acceleration due to vertical gust.

since a safety factor greater than one will guarantee the wing structural performance. However, the wing spar structural failure due to the dynamic load (i.e., fatigue) happen at some point in future.

6.5.5   AEROELASTIC MODEL

The aeroelasticity analysis is often based on flexible aircraft models consisting of traditional rigid-body equations, and a series of second-order differential equations representing the structural modes. In aeroelastic modeling, beam elements (e.g., Euler–Bernoulli beam elements) have traditionally been utilized [35] to represent the wing structure. To establish an aeroelastic model of the clean wing, the linear finite element representations consisting of a lot of small panels (Figure 6.8) may be employed. The wing structure is simulated by beam frames, whose stiffness and mass distributions include the constrains of static aeroelasticity and flutter. The number of state variables (unknowns) is equal to the number of panel, since each panel has a unique displacement (ξ).

The Laplace transform of the open-loop aeroelastic equation of motion [32] in generalized coordinates, excited by control surface motion and atmospheric gusts, is

$$\left(M_{hh}s^2 + B_{hh}s + K_{hh} + qQ_{hh}(s)\right)\xi(s)$$
$$= -\left(M_{hc}s^2 + qQ_{hc}(s)\right)\delta_A(s) - \frac{q}{V}Q_{hG}(s)W_g(s), \qquad (6.32)$$

where V is the air velocity and q is the dynamic pressure. The Q_{hh}, Q_{hc}, Q_{hG} are lift coefficient matrices and should be calculated at various reduced-frequency values. The right-hand side matrices represent the external aerodynamic forces due to control surface commanded deflections (δ_A) and gust velocity vector (W_g). The left-hand side coefficient matrices are the mass,

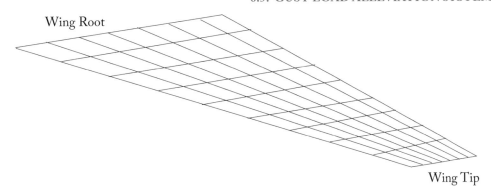

Figure 6.8: Schematic drawing of a structural beam-type model of a clean wing.

stiffness, and lift coefficient matrices associated with structural displacements or elastic deformation (ξ). Hence, two transfer functions of elastic-displacement-to-vertical-gust-velocity and elastic-displacement-to-aileron-deflection are:

$$\frac{\xi(s)}{W_g(s)} = \frac{-\frac{q}{V}Q_{hG}(s)}{M_{hh}s^2 + B_{hh}s + K_{hh} + qQ_{hh}(s)} \tag{6.33}$$

$$\frac{\xi(s)}{\delta_A(s)} = \frac{-\left(M_{hc}s^2 + qQ_{hc}(s)\right)}{M_{hh}s^2 + B_{hh}s + K_{hh} + qQ_{hh}(s)}. \tag{6.34}$$

Modal response analysis to sinusoidal gust excitation can be performed by replacing s in Equation (6.32) by $i\omega$, where ω is the frequency of oscillation, excited by gust velocity inputs. The analysis, simulation, and design processes can be implemented in the time-domain, when Equation (6.33) is reformulated to state-space model.

The vertical discrete gust velocity is defined [29] by:

$$W_g(t) = \frac{1}{2}W_{g\text{max}}\left(1 - \cos\left(\frac{2\pi t}{L_g}\right)\right), \tag{6.35}$$

where Wg_{max} is the maximal gust velocity, (e.g., 1 m/sec), and L_g is the gust length, in terms of the time for a point in the aircraft to fly through it. The gust length, L_g is nondimensionalized with respect to the time for a point in the aircraft to travel across the gust field. The length of a one-minus-cosine vertical gust velocity profile—which yields the maximal wing-root bending moment—should be identified.

6.5.6   LIFT DISTRIBUTION

Dynamic loads at extreme gust cases are often the primary purpose of the gust response analysis. A 3D unsteady aerodynamic panel code may be used to obtain the aerodynamic forces/loads.

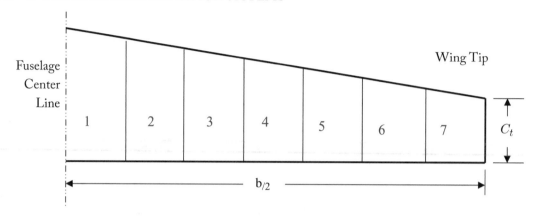

Figure 6.9: Dividing a wing into a number of sections (here, seven) for aerodynamic model.

The lifting surfaces can be discretized using source and doublet panels, and the wake elements are represented by doublet panels. Another linear technique is lifting line theory [33] at which the wing is divided into n segments. This theory is initially developed to determine the lift distribution over the wing by dividing a wing into n segments (Figure 6.9). However, it can be modified to include the effect of a wingtip gust on the distribution of lift across the wing.

The main governing equation of the lifting line theory is:

$$\mu \left(\alpha_o - \alpha \right) = \sum_{n=1}^{N} A_n \sin(n\theta) \left(1 + \frac{\mu n}{\sin(\theta)} \right), \tag{6.36}$$

where α segment's angle of attack; and α_o segment's zero-lift angle of attack and the primary unknowns are A_1 to A_n. The θ is unique for each wing segment, and is found from the line between the lift distribution to the y-axis (Figure 6.10). The parameter μ is defined as:

$$\mu = \frac{\overline{C}_i \cdot C_{l_\alpha}}{4b}, \tag{6.37}$$

where $\overline{C}_i$ denotes the segment's mean geometric chord, C_{l_α} is the segment's lift curve slope in 1/rad, and "b" is the wing span. Each wing segment's lift coefficient is determined using the following equation:

$$C_{L_i} = \frac{4b}{\overline{C}_i} \sum A_n \sin(n\theta). \tag{6.38}$$

The gust contribution to this equation is modeled and is included as an induced angle of attack as indicated in Equation (6.34).

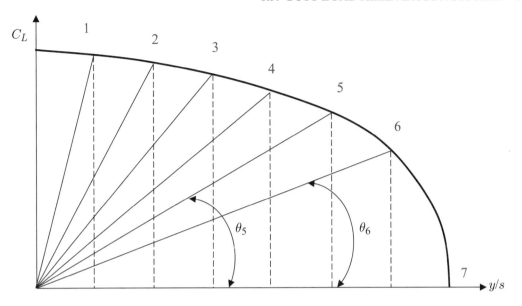

Figure 6.10: Angles corresponding to each segment in lifting-line theory.

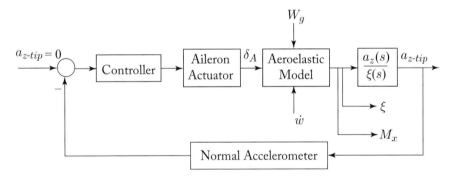

Figure 6.11: Block diagram of a basic gust load alleviation system.

6.5.7   BLOCK DIAGRAM

Figure 6.11 illustrates the block diagram of a basic gust load alleviation system, where the controller sends its command to ailerons (via a hydraulic actuator). The measurement device for this system is an accelerometer installed at the wing tip to measure the vertical acceleration. The acceleration signal should be fast enough for significant alleviation of the loads peak. The system has three inputs and three outputs. Three inputs are the aileron deflection, the gust velocity W_g, and gust linear acceleration ($\dot{W}_g$). Three outputs are displacements (ξ), wing-root vertical acceleration (a_{z-tip}), and the wing-root bending moment (M_x).

The wing aeroelastic model is a structural beam-type model can be generated by software packages such as Autodesk NASTRAN [34]. As the number of structural states is increased (say 50 or 1000), the simulation outputs tend to be more accurate, and more reliable.

The control law may be selected to be based on a low-pass filter for easy and robust application. An effective controller must react sufficiently before the peak of the wing-root bending moment. Any disturbance in roll angle (ϕ) should be damped, so the bank angle will go to zero. The single input to the controller is from a wing-tip accelerometer feedback. Moreover, designer should make sure that aeroservoelastic instability is not induced (by ailerons) when the gust-alleviation loop is closed. In assessing the performance of the control system, the wing section loads, such as shear force, bending moment and torsion, should be examined.

For easy and robust application, a controller (K) based on simple low-pass filter is recommended.

$$K(s) = \frac{k_c}{\tau_c s + 1}. \tag{6.39}$$

Reference [32] found that the best alleviation which provides acceptable stability margins for a wind tunnel model application is obtained with $k_c = 0.09$ sec^2/m and $\tau_c = 0.56$ sec. Control surfaces should be controlled (i.e., deflected) in a way to significantly alleviate the extreme wing-root bending moments at nominal flight velocities for extreme wing-tip accelerations. The bending moment at the root part of the wing is alleviated by the aileron deflection. To alleviate gust, some Airbus A320 are using spoilers, while Airbus 380 is employing ailerons.

A typical time histories of the GLAS-off and GLAS-on (i.e., open- and closed-loop) wing root bending moment in response to the discrete gust excitation are compared in Figure 6.12. As observed, the bending moment can be considerably reduced with the gust allevia-

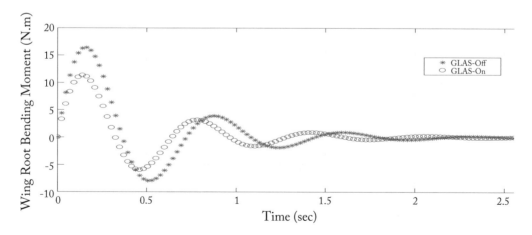

Figure 6.12: A typical time histories of the GLAS-off and GLAS-on wing root bending moment.

Figure 6.13: Embraer ERJ-190 Lineage (courtesy of Mark Harkin).

Figure 6.14: Embraer ERJ-190 Lineage 1000 Cockpit (courtesy of Sergey Ryabtsev).

tion. Furthermore, the GLAS can affect the wing's natural frequency of vibration, consequently having great effects on the flutter characteristics of the wing.

A short- to medium-range ultra-large business jet aircraft Embraer ERJ-190 Lineage that is equipped with a GLAS is shown in Figure 6.13. The aircraft has a wing area of 92.5 m², a maximum takeoff mass of 54,500 kg, and is equipped with two turbofan engines. The aircraft can travel a distance of 8,500 km with a cruise speed of 472 knot, and has a service ceiling of 41,000 ft. The cockpit of this aircraft is depicted in Figure 6.14. The flight deck has five displays for crews (including one primary display for each pilot and copilot). Both pilot and copilot have access to the autopilot interface to select and engage various AFCS modes.

6.6   QUESTIONS

6.1. Name at least three CASs.

6.2. What control surfaces are used by Airbus 380 to alleviate the gust loads?

6.3. Briefly describe the function of a normal acceleration CAS.

6.4. Briefly describe the function of a pitch rate CAS.

6.5. Briefly describe the function of a lateral-directional CAS.

6.6. Briefly describe the function of a gust load alleviation system.

6.7. Briefly describe the function of a relaxed static stability system.

6.8. Define load factor.

6.9. What title is employed by Ref. [3] to refer to a CAS?

6.10. What is the best aircraft location for a normal accelerometer?

6.11. List primary elements of a basic block diagram of a normal acceleration CAS.

6.12. List primary elements of a basic block diagram of a pitch rate CAS.

6.13. List primary elements of a basic block diagram of a lateral-directional CAS.

6.14. List primary elements of a basic block diagram of a gust load alleviation system.

6.15. Name a few flight operations and maneuvers that require a well-behaved pitch rate response.

6.16. What is C-star parameter?

6.17. What parameters affect the pitching moment due to inertia coupling?

6.18. What is the main objective of an aileron-rudder interconnect?

6.19. What are the lateral acceleration and normal acceleration, and how they are measured?

6.20. Briefly describe how the aircraft aeroelastic is modeled.

6.21. Introduce the discrete gust idealization.

6.22. How is the wing root bending moment due to a wing tip vertical gust determined?

6.23. How is the induced angle of attack due to a wing tip vertical gust determined?

6.24. Write the format of an elastic-displacement-to-vertical-gust-velocity transfer function.

6.25. Discuss the lifting line theory in brief.

6.26. Draw the block diagram of a basic gust load alleviation system.

References

[1] Stevens, B. L., Lewis, F. L., and Johnson, E. N., *Aircraft Control and Simulation*, 3rd ed., John Wiley, 2016. DOI: 10.1108/aeat.2004.12776eae.001. 10, 14, 16, 50, 57, 66, 112, 133, 138

[2] Roskam, J., Airplane flight dynamics and automatic flight controls, *DARCO*, 2007. 9, 50, 51, 53, 56, 60, 66, 73, 114, 121, 122, 131

[3] McLean, D., *Automatic Flight Control Systems*, Prentice Hall, 1990. DOI: 10.1177/002029400303600602. 60, 64, 133, 137, 138, 152

[4] Blakelock, J. H., *Automatic Control of Aircraft and Missiles*, 2nd ed., John Wiley, 1991. 120

[5] Nelson, R., *Flight Stability and Automatic Control*, McGraw Hill, 1989. 11, 53, 63, 64, 118

[6] Etkin, B. and Reid, L. D., *Dynamics of Flight, Stability and Control*, 3rd ed., Wiley, 1996. DOI: 10.1063/1.3060977. 53

[7] Siouris, G. M., *Missile Guidance and Control Systems*, Springer, 2004. DOI: 10.1115/1.1849174.

[8] Hoak, D. E., Ellison, D. E., et al., USAF stability and control DATCOM, *Flight Control Division*, Air Force Flight Dynamics Laboratory, Wright-Patterson AFB, OH, 1978.

[9] MIL-STD-1797, Flying qualities of piloted aircraft, Department of Defense, Washington DC, 1997. 36, 60

[10] MIL-F-8785C, Military specification: Flying qualities of piloted airplanes, Department of Defense, Washington DC, 1980. 36, 121

[11] Sadraey, M., *Aircraft Performance Analysis: An Engineering Approach*, CRC Press, 2017. 98, 113

[12] https://www.mathworks.com/help/control/examples/yaw-damper-design-for-a-747-jet-aircraft.html 119

[13] Duncan, J. S., *Pilot's Handbook of Aeronautical Knowledge*, U.S. Department of Transportation, Federal Aviation Administration, Flight Standards Service, 2016. 75, 76, 77

[14] Parkinson, B. W., O'Connor, M. L., and Fitzgibbon, K. T., *Global Positioning System: Theory and Applications*, Chapter 14: Aircraft Automatic Approach and Landing Using GPS, American Institute of Aeronautics and Astronautics, 1996. DOI: 10.2514/4.472497. 89

[15] Goold, I., Boeing forges ahead with flight-test campaigns, *AIN (Aviation International News)*, online, November 8, 2017. 92

[16] Jayakrishnan, S. and Harikumar, R., Trajectory generation on approach and landing for a RLV using NOC approach, *International Journal of Electrical, Electronics and Data Communication*, 1(6), August 2013. 88

[17] *Advanced Avionics Handbook*, Federal Aviation Administration, FAA-H-8083–6, U.S. Department of Transportation, Washington DC, 2009.

[18] Dods, J. B., Jr. and Tinling, B. E., Summary of results of a wind-tunnel investigation of nine related horizontal tails, *Technical Note 3497, NACA*, 1955. 131

[19] Brown, J. A., Stall avoidance system for aircraft, Patent Number: 4,590,475, United States Patent, May 20, 1986. 126

[20] Malaek, S. M. and Kosari, A. R., Novel minimum time trajectory planning in terrain following flights, *IEEE Transactions on Aerospace and Electronic Systems*, 43(1), January 2007. DOI: 10.1109/taes.2007.357150. 101

[21] Sadraey, M., *Unmanned Aerial Systems Design*, Wiley, 2019. 103, 105

[22] Capello, E., Guglieri, G., and Quagliotti, F., A waypoint-based guidance algorithm for mini UAVs, *2nd IFAC Workshop on Research, Education and Development of Unmanned Aerial Systems*, Compiegne, France, November 20–22, 2013. DOI: 10.3182/20131120-3-fr-4045.00005. 104

[23] Bischoff, D. E., Development of longitudinal equivalent system models for selected U.S. navy tactical aircraft, *Report No. NADC-81069–60, Aircraft and Crew Systems Technology Directorate*, Naval Air Development Center, PA, 1981. DOI: 10.21236/ada109488. 116

[24] Hoblit, F. M., Gust loads on aircraft: Concepts and applications, *AIAA*, Washington, DC, 1988. DOI: 10.2514/4.861888. 142

[25] Karpel, M., Moulin, B., Anguita, L., Maderuelo, C., and Climent, H., Aeroservoelastic gust response analysis for the design of transport aircrafts, *AIAA Paper 2004–1592*, April 2004. DOI: 10.2514/6.2004-1592. 143

[26] Devlin, B. T. and Girts, R. D., MD-11 automatic flight system, *Proc. of IEEE/AIAA 11th Digital Avionics Systems Conference*, pages 174–177, 1992. DOI: 10.1109/dasc.1992.282164. 12

[27] Regional Business News, Garmin flies G5000 in a Beechjet 400A Airline Industry Information, September 24, 2014. 117

[28] Napolitano, M., *Aircraft Dynamics: From Modeling to Simulation*, Wiley, 2011. 144

[29] Federal Aviation Regulations, Part 23, Airworthiness Standards: Normal, Utility, Aerobatic, and Commuter Category Airplanes, Federal Aviation Administration, Department of Transportation, Washington DC. 42, 144, 147

[30] Federal Aviation Regulations, Part 25, Airworthiness Standards: Transport Category Airplanes, Federal Aviation Administration, Department of Transportation, Washington DC. 144

[31] Karpel, M., Moulin, B., and Chen, P. C., Dynamic response of aeroservoelastic systems to gust excitation, *Journal of Aircraft*, 42(5):1264–1272, 2005. DOI: 10.2514/1.6678. 143

[32] Moulin, B. and Karpel, M., Gust loads alleviation using special control surfaces, *Journal of Aircraft*, B.44(1), January–February 2007. DOI: 10.2514/1.19876. 142, 146, 150

[33] Houghton, E. L. and Carpenter, P. W., *Aerodynamics for Engineering Students, Butterworth–Heinemann*, 7th ed., 2016. 148

[34] Autodesk Nastran In-CAD 2019, *ASCENT*, Center for Technical Knowledge, 2018. 150

[35] Haghighat, S., Liu, H., and Martins, J., Model-predictive gust load alleviation controller for a highly flexible aircraft, *Journal of Guidance, Control and Dynamics*, 35(6):1751–1766, 2012. DOI: 10.2514/1.57013. 146

[36] Prats, X., et al., Requirements, issues, and challenges for sense and avoid in unmanned aircraft systems, *Journal of Aircraft*, 49(3), May 2012. DOI: 10.2514/1.c031606. 104

[37] Angelov, P., *Sense and Avoid in UAS: Research and Applications*, Wiley, 2012. DOI: 10.1002/9781119964049. 104

[38] Dorf, R. C. and Bishop, R. H., *Modern Control Systems*, 13th ed., Pearson, 2017. DOI: 10.1109/tsmc.1981.4308749. 7, 25, 37, 41, 61

[39] Ogata, K., *Modern Control Engineering*, 5th ed., Prentice Hall, 2010. DOI: 10.1115/1.3426465. 7, 25, 37, 39

[40] Anderson, B. D. O. and Moore, J. B., *Optimal Control: Linear Quadratic Methods*, Dover, 2007. 37

[41] Kendoul, F., Survey of advances in guidance, navigation, and control of unmanned rotorcraft systems, *Journal of Field Robotics*, 29(2):315–378, 2012. DOI: 10.1002/rob.20414. 4

[42] Menon, P. K. A., Nonlinear command augmentation system for a high performance aircraft, *AIAA-93–3777-CP, American Institute of Aeronautics and Astronautics, Guidance, Navigation and Control Conference*, Monterey, CA, August 9–11, 1993. DOI: 10.2514/6.1993-3777. 134

Author's Biography

MOHAMMAD SADRAEY

Dr. Mohammad Sadraey is an Associate Professor in the College of Engineering, Technology, and Aeronautics at the Southern New Hampshire University, Manchester, New Hampshire, and the national Vice President of Sigma Gamma Tau honor society in USA. Dr. Sadraey's main research interests are in aircraft design techniques, aircraft performance, flight dynamics, autopilot, and design and automatic control of unmanned air vehicles. He received his M.Sc. in Aerospace Engineering in 1995 from RMIT, Australia, and his Ph.D. in Aerospace Engineering from the University of Kansas, USA in 2006. Dr. Sadraey is a senior member of the American Institute of Aeronautics and Astronautics (AIAA), and a member of American Society for Engineering Education (ASEE), and is in Who's Who in America for many years. He has over 24 years of professional experience in academia and industry. Dr. Sadraey is the author of six other books including *Aircraft Design: A Systems Engineering Approach*, *Design of Unmanned Aerial Systems*, published by Wiley Publications in 2012, and 2019, and *Aircraft Performance Analysis* by CRC in 2016.